Mangrove management a

Mangrove management and conservation: present and future

Edited by Marta Vannucci

United Nations
University Press
TOKYO · NEW YORK · PARIS

© United Nations University, 2004

The views expressed in this publication are those of the authors and do not necessarily reflect the views of the United Nations University.

United Nations University Press
United Nations University, 53-70, Jingumae 5-chome,
Shibuya-ku, Tokyo, 150-8925, Japan
Tel: +81-3-3499-2811   Fax: +81-3-3406-7345
E-mail: sales@hq.unu.edu   (general enquiries): press@hq.unu.edu
http://www.unu.edu

United Nations University Office in North America
2 United Nations Plaza, Room DC2-2062, New York, NY 10017, USA
Tel: +1-212-963-6387   Fax: +1-212-371-9454
E-mail: unuona@ony.unu.edu

United Nations University Press is the publishing division of the United Nations University.

Cover design by Joyce C. Weston
Cover photograph by Shigeyuki Baba

Printed in the United States of America

UNUP-1084
ISBN 92-808-1084-7

Library of Congress Cataloging-in-Publication Data

Mangrove management and conservation : present and future / edited by Marta Vannucci.
   p. cm.
Includes bibliographical references and index.
ISBN 9280810847 (pbk.)
1. Mangrove forests—Management. 2. Mangrove conservation. I. Vannucci, M. (Marta), 1921–
SD397.M25M373   2004
634.9′73763—dc22   2004001198

Contents

List of tables, photographs, and figures ix

Note on measurements.. xvii

Foreword... xix

Introduction and keynote presentation.............................. 1

Introduction... 3
    Marta Vannucci

Keynote presentation: What we can do for mangroves 8
    S. Baba

Part I The mangrove ecosystem: Structure and function 37

1 Some ecological aspects of the morphology of
   pneumatophores of *Sonneratia alba* and *Avicennia officinalis*... 39
   T. Nakamura, R. Minagawa, and S. Havanond

2 Introduction of *Sonneratia* species to Guangdong Province,
   China 45
   Zhongyi Chen, Ruijiang Wang, and Zebin Miao

3 Research into, and conservation of, mangrove ecosystems in
   Indonesia... 51
   Aprilani Soegiarto

4 Status of Indian mangroves: Pollution status of the
   Pichavaram mangrove area, south-east coast of India............ 59
   AN Subramanian

5 The role of aquatic animals in mangrove ecosystems............ 76
   Shigemitsu Shokita

6 Effects of mangrove restoration and conservation on the
   biodiversity and environment in Can Gio District............... 111
   Phan Nguyen Hong

7 Below-ground carbon sequestration of mangrove forests in the
   Asia-Pacific region.. 138
   Kiyoshi Fujimoto

Part II Function and management **147**

8 Sustainable use and conservation of mangrove forest
   resources with emphasis on policy and management practices
   in Thailand.. 149
   Sanit Aksornkoae

9 Role of the national government in the economic
   development of the mangroves of Fiji............................. 161
   Mesake Senibulu

10 Conflicting interests in the use of mangrove resources in
   Pakistan 167
   Mohammad Tahir Qureshi

11 Mangroves, an area of conflict between cattle ranchers and
   fishermen.. 181
   Patricia Moreno-Casasola

12 Philippine mangroves: Status, threats, and sustainable
   development 192
   J.H. Primavera

13	Co-management of coastal fisheries resources in tropical and subtropical regions... *Shinichiro Kakuma*	208
14	Mangrove Rehabilitation and Coastal Resource Management Project of Mabini–Candijay, Bohol, Philippines: Cogtong Bay *Robert S. Pomeroy and Brenda M. Katon*	219

Part III Uses and policies.. 231

15	Towards sustainable use and management for mangrove conservation in Viet Nam .... *Motohiko Kogo and Kiyomi Kogo*	233
16	Mangrove forestry research in Bangladesh....................... *A.F.M. Akhtaruzzaman*	249
17	Socio-economic study of the utilization of mangrove forests in South-East Asia... *Kazuhiro Ajiki*	257
18	Sustainable mangrove management in Indonesia: Case study on mangrove planting and aquaculture.......................... *Atsuo Ida*	270
19	Sustainable use and conservation management of mangroves in Zanzibar, Tanzania .... *Thabit S. Masoud and Robert G. Wild*	280

Part IV Summary of presentations and guidelines, and action plan 295

20	Summary of presentations and guidelines for future action..... *Marta Vannucci and Zafar Adeel*	297
21	Mangroves action plan .... *Zafar Adeel*	302

Acronyms 306

Contributors 308

Index 311

List of tables, photographs, and figures

Table 2.1	Mangrove species of Guangdong Province	46
Table 2.2	Height of *Sonneratia apetala* after planting	49
Table 4.1	Area under mangroves in India	61
Table 4.2	Physical and chemical parameters observed in the Pichavaram mangroves during June 1999 and October 1999	68
Table 5.1	Macrofaunal community at the mangal areas of South-East Asia and Okinawa	78
Table 5.2	Macrofaunal habitats and representative animals on Okinawan and Thai mangals	88
Table 5.3	Comparison of species composition and abundance of tree fauna between Okinawan and Thai mangals	90
Table 5.4	Crustacean species found at the mangrove swamp of the Okukubi River in 1979 and 1997, showing feeding habits and habitat	92
Table 5.5	Average weekly rates of leaf consumption by *Helice leachi* per gram body weight when provided with green, yellow, and brown leaves of *Bruguiera gymnorrhiza* separately	96
Table 5.6	Average weekly rates of leaf consumption by *Helice leachi* per gram body weight when	

	provided with green, yellow, and brown leaves of *Bruguiera gymnorrhiza* together	96
Table 5.7	Nutritional composition of green, yellow, and brown leaves of *Bruguiera gymnorrhiza* per 100 g wet weight	97
Table 5.8	Nutritional composition of green, yellow, and brown leaves of *Bruguiera gymnorrhiza* per 100 g dry weight	97
Table 6.1	Can Gio mangroves in 24 forestry units	117
Table 6.2	Rare animals in Can Gio	123
Table 6.3	Physicochemical characteristics of mangrove soils after reforestation....................................	125
Appendix 6.1	Mangrove flora in Can Gio Mangrove Biosphere Reserve	127
Appendix 6.2	Zoobenthos in Can Gio Mangrove Biosphere Reserve	128
Appendix 6.3	Fish in Can Gio Mangrove Biosphere Reserve.....	130
Appendix 6.4	Amphibians and reptiles in Can Gio Mangrove Biosphere Reserve	132
Appendix 6.5	Birds in Can Gio Mangrove Biosphere Reserve ...	133
Appendix 6.6	Mammals in Can Gio Mangrove Biosphere Reserve	135
Table 7.1	Below-ground carbon storage in mangrove forests in the Asia-Pacific region	141
Table 7.2	Carbon-burial rates of mangrove forests in the Asia-Pacific region	142
Table 8.1	Changes in the existing mangrove forest in the period 1961–1996	152
Table 8.2	Area of mangrove land-use zones by province.....	154
Table 9.1	Principal species of Fijian mangrove vegetation....	163
Table 10.1	Mangrove species of Pakistan, with their occurrence...	169
Table 10.2	Indus River average annual (and seasonal) discharge volumes downstream of Kotri barrage...	169
Table 10.3	Percentage salinity in the Indus delta during 1996...	170
Table 10.4	Frequent use of mangroves for fodder..............	172
Table 10.5	Monthly consumption of mangroves for fuel.......	173
Table 10.6	Estimated number of camels in 1996................	175
Table 10.7	Rehabilitation of mangroves in the Indus delta and Balochistan coast.................................	177
Table 11.1	Stakeholders in the area.............................	185

Table 12.1	Major and minor mangrove species in the Philippines	193
Table 12.2	Economic values placed on products and services of mangrove systems	195
Table 12.3	Total mangrove and brackish-water culture-pond area in the Philippines	197
Table 12.4	Some Philippine regulations on fish-ponds and mangrove conversion	200
Table 12.5	Some Philippine regulations on mangrove conservation and rehabilitation	202
Table 13.1	Major fish species caught in Taketomi District	217
Table 15.1	Reforestation activities of the ACTMANG project in 1994–1999	237
Table 15.2	Main activities regarding education	242
Table 16.1	Economically important plants of the Sundarbans and their uses	252
Table 17.1	Occupational composition in Barangay Lincod, 1992	259
Table 17.2	Household composition and occupations in Tam Thon Hiep Commune, 1994	262
Table 17.3	Household composition and occupations in Ban Tha Pradu, 1997	266
Table 18.1	Examples of mangrove planting and traditional extensive aquaculture	276
Table 19.1	Forest areas of Zanzibar	282
Table 19.2	Forest area and standing volumes for Charawe and Ukongoroni	291

NB: All photographs can be found at the end of the chapter to which they refer.

Photo K.1	Small mangrove trees (*Kandelia candel*) in Naha City, Okinawa, Japan	16
Photo K.2	Small mangrove trees in Naha City, Okinawa	16
Photo K.3	Mangroves on Iriomote Island, Okinawa	17
Photo K.4	About 200 ha of mangrove forests along the Nakama River, Iriomote Island, Okinawa	17
Photo K.5	Buttress roots of a very old specimen of *Heritiera littoralis*, Iriomote Island, Okinawa	18
Photo K.6	*Rhizophora* sp. with very large stilt roots, Esmeraldas, Ecuador	18

xii LIST OF TABLES AND FIGURES

Photo K.7	A child in white clothing stands near the trunk of a very large tree (*Sonneratia* sp.), Indonesia	19
Photo K.8	Destruction of mangrove forests during construction of a resort hotel.......................	19
Photo K.9	Abandoned shrimp ponds, Bali Island, Indonesia..	20
Photo K.10	Destruction of mangrove forests during construction of shrimp ponds, Ecuador............	20
Photo K.11	Leaves and young branches of *Avicennia* sp. are an important source of cattle fodder in Iran........	21
Photo K.12	Water buffaloes in Indonesia prefer to eat the leaves and young twigs of *Avicennia marina*	21
Photo K.13	Fishes gather under mangrove trees on Iriomote Island, Okinawa.......................................	22
Photo K.14	Birds gather on mangrove trees in Florida, USA ..	22
Photo K.15	Monkeys in mangroves in Viet Nam; many more were present, although not visible here............	23
Photo K.16	Sea lions taking an afternoon nap in the shade of *Avicennia* trees on the Galapagos Islands, Ecuador...	23
Photo K.17	Fronds of nipa palms (*Nypa fruticans*) provide good materials for thatching in Viet Nam	24
Photo K.18	Thatching with nipa palm leaves, Viet Nam........	24
Photo K.19	Artistic thatching with fronds of nipa palm, Viet Nam..	25
Photo K.20	Mangrove timber being used as firewood in Fiji ...	25
Photo K.21	Bark being peeled off wooden billets that will be burned in kilns to make charcoal, Thailand	26
Photo K.22	Charcoal kiln, Thailand.............................	26
Photo K.23	Dyes are extracted from the bark of *Bruguiera* and *Rhizophora* spp. on Iriomote Island, Okinawa..	27
Photo K.24	Dying cloth with extracts from the bark of *Bruguiera gymnorrhiza* and *Rhizophora stylosa*, Iriomote Island, Okinawa	27
Photo K.25	Poles of *Bruguiera* and *Rhizophora* spp. are used in construction work	28
Photo K.26	Abandoned shrimp-ponds on Bali Island, Indonesia	28
Photo K.27	Potted seedlings of *Avicennia* sp., Bali..............	29
Photo K.28	Seeds of *Sonneratia alba*.............................	29
Photo K.29	Propagules of *Rhizophora apiculata*, Bali	30
Photo K.30	Propagules of *Rhizophora mucronata*, Bali.........	30
Photo K.31	Mangrove nursery, Bali.............................	31

LIST OF TABLES AND FIGURES xiii

Photo K.32	Experimental mangrove plantation in an abandoned shrimp pond, Bali	31
Photo K.33	Dense mangrove plantation that will require thinning, Indonesia...................................	32
Photo K.34	Japanese ladies planting mangroves on Lombok Island, Indonesia	32
Photo K.35	Local inhabitants and Japanese volunteers planting mangroves on Fiji	33
Photo K.36	Mangrove plantation on the occasion of the Fourth General Assembly of the International Society for Mangrove Ecosystems (ISME), September 1999, Bali, Indonesia	33
Photo K.37	Mangrove plantation about 20 years old, Viet Nam	34
Photo K.38	Mangrove plantation about 15 years old, Viet Nam	34
Photo K.39	Mangrove plantation about 15 years old, Thailand	35
Photo K.40	Pohnape Island in Micronesia, like other oceanic islands, faces the danger of a rise in mean sea level (Photo Nakamura)	35
Photo 5.1	Crustaceans found in Okinawan mangals (I): (A) *Thalassina anomala*; (B) *Mictyris brevidactylus*; (C) ecdysis of *Scylla serrata*; (D) *Uca (Amphiuca) chlorophthalma crassipes*; (E) *Helice leachi*; (F) mating of *Neoepisesarma* (N.) *lafondi* under laboratory conditions	105
Photo 5.2	Crustaceans found in Okinawan mangals (II): (A) ovigerous *Clibanarius longitarsus*; (B) ovigerous *Macrobrachium scabriculum* living at the boundary between brackish and fresh water; (C) *Metapenaeus moyebi*; (D) *Penaeus monodon*.......	106
Photo 5.3	Molluscs found in Okinawan and Thai mangals: (A) *Geloina coaxans*, muddy channel, Iriomote Island; (B) clusters of *Terebralia palustris*, rasping fallen mangrove leaves, Ao Khung Kraben mangal, Chanthaburi Province of eastern Thailand; (C) *Telescopium telescopium* on the mud-flat of Ranong mangal, Thailand; (D) *Rhizophomurex capucinus* preying on the young of *Terebralia palustris*, Ao Khung Kraben mangal; (E) *Cerithidea quadratai* hanging from a dead mangrove branch, Ao Khung Kraben mangal; (F)	

	Elobium auris-judae at rest on the trunk of *Rhizophora* sp.; (G) a cluster of *Cerithidea rhizophorarum* at rest on the trunk of *Rhizophora stylosa*, Ishigaki Island mangal, Okinawa...........	107
Photo 5.4	Tree-dwelling fauna on Ranong mangal, Thailand: (A) *Aegiceras corniculatum* at the seaward edge of transect 2, Ranong, with sedentary species such as *Melina ephippium*, *Saccostrea cucullata*, and *Balanus amphitrite*; (B) a large colony of *M. ephippium* attached by its byssus to the trunk of *A. corniculatum*; (C) *Balanus amphitrite* attached to the trunk and branches of *Sonneratia alba*; (D) *S. cucullata* on the prop roots of *Rhizophora mucronata* growing on the creek side of transect 3; (E) *Cymia gralata* preying on the *S. cucullata* encrusting the roots of *A. corniculatum*	108
Photo 5.5	Fish species found in Okinawan mangals (I): (A) *Carcharinus leucus* (also shown here are *Mugil cephalus*, *Platycephalus* sp., and *Acanthopagrus* sp.); (B) *Evenchelys macrurus*; (C) *Mugil cephalus*; (D) *Sphyraena barracuda*; (E) *Apogon amboinensis*; (F) *Caranx sexfasciatus*	109
Photo 5.6	Fish species found in Okinawan mangals (II): (A) *Gerres filamentosus*; (B) *Monodactylus argenteus*; (C) *Terapon jarbua*; (D) *Acanthopagrus silvicolus*; (E) *Siganus guttatus*; (F) *Ophiocara poracephala* ..	110
Photo 10.1	*Rhizophora mucronata*: assisted natural regeneration in the Indus delta.....................	180
Photo 10.2	*Avicennia marina*: plantation in trench system in Miani Hor	180
Photo 15.1	*Kandelia candel* reforestation in Tinh Gio: the seedlings are 6 months old.........................	247
Photo 15.2	Shrimp farming combined with the planting of *Rhizophora apiculata* at Ben Tre in the Mekong delta...	247
Photo 15.3	Five lodges have been established in Can Gio, Ho Chi Minh City, for education and research regarding mangroves................................	248
Figure K.1	What we can do for mangroves: Goals	12
Figure K.2	Means of accomplishing the goals outlined in figure K.1	12

LIST OF TABLES AND FIGURES xv

Figure K.3	Roles of the International Society for Mangrove Ecosystems (ISME)	13
Figure 1.1	Numbers and types of pneumatophores at various distances from (a) large, (b) medium-sized, and (c) small (young) trees of *Sonneratia alba*	41
Figure 1.2	Proportions of various types of pneumatophores found on the roots of trees of *Sonneratia alba*	42
Figure 1.3	Comparison of proportions of pneumatophores of types A–D on roots of *Sonneratia alba* from Ranong and Phang-nga	43
Figure 1.4	Correlation between young tree height and cable-root thickness in *Avicennia officinalis* and *A. alba*	43
Figure 2.1	Mean growth height of *Sonneratia apetala* at Leizhou, Furongwan, and Jiepao at various times after planting	49
Figure 4.1	Statewise distribution of mangroves in India	62
Figure 4.2	Pichavaram mangrove forests on the south-east coast of India	65
Figure 4.3	Spatial distribution of γ-HCH in the Pichavaram mangrove and Vellar estuary during the dry and wet seasons	71
Figure 4.4	Spatial distribution of DDT in the Pichavaram mangrove and Vellar estuary during the dry and wet seasons	71
Figure 4.5	Schematic representation of the flux of γ-HCH in the Vellar watershed	72
Figure 5.1	Sketch of Thailand, showing location of study sites (and transects T1–T9) at Smare Kaow and Ranong	80
Figure 5.2	Zonation and abundance of macrofauna at Smare Kaow	81
Figure 5.3	Zonation and abundance of macrofauna in transect 1 at Ranong	83
Figure 5.4	Zonation and abundance of macrofauna in transect 3 at Ranong	84
Figure 5.5	Zonation and abundance of macrofauna in transect 6 at Ranong	86
Figure 5.6	Four biotopes of the mangrove ecosystem, and the zonation of dominant groups and species of the macrofauna in Thailand	87
Figure 5.7	Feeding habits of estuarine fishes	99
Figure 5.8	Food chains in mangroves of Iriomote Island, Okinawa	101

Figure 6.1	Map of the Can Gio District, Ho Chi Minh City...	112
Figure 6.2	Map of Can Gio mangrove forest and forestry units..	118
Figure 7.1	Relationship between carbon-burial period and stored carbon for *Rhizophora* forest in the Asia-Pacific region	144
Figure 8.1	Sketch of Thailand, showing the main mangrove areas	159
Figure 12.1	Change in area of mangroves and brackish-water culture ponds in the Philippines, 1951–1988/1990..	198
Figure 12.2	Parallel trends in declining mangrove area and municipal fisheries production, and increasing brackish-water pond area and aquaculture production in the Philippines, 1976–1998...........	199
Figure 12.3	Guidelines to mangrove zones as sites for (extensive) aquaculture ponds......................	201
Figure 13.1	Catch transition of bottom fish, fish aggregating device (FAD) fish, and *Thysanoteuthis rhombus* (giant squid)..	209
Figure 13.2	Catch transition of shellfish and sea urchin.........	210
Figure 13.3	The fisheries extension process in Samoan villages...	213
Figure 13.4	The fisheries extension process in Onna Village, Okinawa...	214
Figure 13.5	Marine reserves in Onna Village....................	215
Figure 13.6	Configurations of Okinawan FADs.................	216
Figure 14.1	Location map of Cogtong Bay	221
Figure 15.1	Sketch of Viet Nam showing project areas	238
Figure 16.1	Coastal belt of Bangladesh showing location of the planted and natural mangrove forests	251
Figure 18.1	Relationships between, and major products of, each component of the project	272
Figure 18.2	Flow chart of nursery work: *Nursery Manual* for mangrove species at Benoa Port, Bali	273
Figure 18.3	Flow chart of silvicultural activities: *Silviculture Manual* for mangroves in Bali and Lombok........	275
Figure 19.1	Forest types and proportions in Zanzibar	282
Figure 19.2	Institutional arrangements for community-based conservation in Zanzibar	289

Note on measurements

In this volume:

1 billion = 1,000 million
1 ton = 2,240 lb
1 tonne = 1 metric ton (1,000 kg)
$1 = 1 US dollar

Foreword

> Man has too long forgotten that the earth was given to him for usufruct alone,
> not for consumption, still less for profligate waste.
> George Perkins Marsh, *Man and Nature*, 1864

Coastal zones are some of the richest areas in terms of natural resources and biological diversity. These are the areas that have also seen fierce human competition for commercial exploitation, urban development, recreational uses, and waste disposal. Underlying this competition is the fact that the vast majority of megacities and fastest-growing cities are located in (or very close to) coastal zones. The trend for rural populations to migrate to coastal urban areas has greatly accelerated in recent years; this is of particular concern in developing countries, where urban areas often have poor infrastructure, planning, and resources, to begin with. Another threat to coastal zones has emerged in the last two decades – global warming and the attendant rise in sea level. Some of the recent scientific predictions on accelerated sea-level rise as a result of anthropogenically induced global warming are, indeed, alarming.

The focus of this book is on the coastal ecosystems of the tropical and subtropical regions. These ecosystems – including coral reefs, mangroves, and estuarine wetlands – are important elements of global primary productivity. It is no surprise that about two-thirds of fish caught worldwide spend some stage of their lives in these ecosystems or are indirectly dependent on them. It is heartening to observe that considerable research work has been undertaken to evaluate the status and threats to coral reefs. Some of the work undertaken by teams from the World Resources Institute and their partners in this respect is highly commendable; however, the same level of intensity in research work and global attention is not as obvious for mangrove ecosystems. Any number of reasons may be cited for this oversight, but ways of improving the situation must be found, considering the importance of these ecosystems in the overall picture.

Mangroves must be viewed as an integrated coastal ecosystem comprising a rich diversity of flora and fauna. These ecosystems contain protective habitats – such as spawning grounds, a nursery for juveniles, and secure feeding grounds – for a wide number of fish, crab, shrimp, and mollusc species. At the same time, these ecosystems serve as a sanctuary for indigenous and migratory bird species. Indeed, the richness of biological diversity in these ecosystems, as described by the broad spectrum of researchers in this book, is impressive.

Like other coastal ecosystems, mangroves are not safe from human intervention and destruction. They were traditionally managed by small coastal communities at a sustainable level, but their intense exploitation has led to an ever-worsening picture. These ecosystems have become an easy target for the extraction of wood for fuel and construction, the production of food, and waste disposal. The interruption of freshwater flow into mangroves as a result of dam construction and river-water diversion has indirectly caused severe – and, in some cases, irreversible – damage. Land-based sources of pollutants and excess nutrients are yet another culprit in the long list of negative factors. One damaging factor that stands above all the rest is unfettered aquaculture and shrimp farming. The areas suitable for mangroves also have ideal conditions for shrimp farming. This, coupled with an incredibly high economic return on shrimp farming, has been the undoing of mangroves: large tracts of former mangrove ecosystems have now been laid waste because of mismanaged shrimp farming, which has made them too rich in nutrients and antibiotics to sustain a thriving ecosystem of any kind.

There is some ambiguity in defining and understanding the "original" or baseline status of mangrove forests the world over; nevertheless, significant information exists that can be (and has been) used to assess the overall level of destruction of mangroves. Some of the more dramatic figures show that about 75 per cent of the mangroves in Myanmar, Malaysia, Pakistan, and the Philippines have been destroyed over the last few decades; somewhat lower destruction figures are cited for a number of other countries – for example, about 50 per cent for Angola, Gabon, Indonesia, and Thailand. Regardless of the complete accuracy of these numbers, it is quite obvious that the scale of destruction is of catastrophic proportions. Against this backdrop, it is interesting to note a lack of focused research and development work on mangrove ecosystems. Even more importantly, there appears to be almost a vacuum in governmental policies dealing with mangroves: ironically (and tragically), most governments list mangrove areas as "wastelands." These deficiencies in scientific and policy arenas have prompted international agencies to react and to take a more proactive stance.

United Nations University (UNU) has always given a high priority to research and capacity building related to coastal ecosystems. The rationale for this prioritization is quite obvious when one considers the anthropogenic impacts as well as the human reliance on these ecosystems. A number of initiatives led by UNU have focused on monitoring pollution, evaluating ecosystem health, and understanding the socio-economic factors related to the health of coastal ecosystems. These efforts have typically engaged multidisciplinary teams investigating the interlinked issues, resulting in networks of researchers and professionals working towards the ultimate goal of integrated coastal management.

The UNU's work on coastal-resources management stems from the early period of the University's existence. As an example, a project on "Coastal-Resources Management" related to the humid tropical coasts in Indonesia was initiated in 1979. It was focused on the Cimanuk delta in northern Java, which is typical of intensively used deltas in Indonesia. The project concentrated on training and research courses and on conducting management workshops on selected coastal environments, particularly in tropical countries. More recently, the UNU has been involved with the Asia-Pacific Mussel Watch, which is an integral part of the International Mussel Watch Programme. This programme aims to provide an assessment of the status and trends of chemical contaminants in the world's coastal waters, utilizing mussels and other sentinel bivalves as indicators of ecosystem health.

Over the past few years, the UNU has been running a project on environmental monitoring and governance in the East Asian coastal hydrosphere. This project has focused on three elements: (1) building partnerships between research groups, international organizations, and individuals working on coastal issues; (2) monitoring the coastal waters to determine pollution levels – particularly those of endocrine-disruptive compounds; and (3) conducting research and training on ways to conserve and protect the region's precious mangrove ecosystems.

It was a timely intervention when the Government of Japan offered its financial support to the UNU to tackle some of the challenging issues related to mangroves. The UNU developed an alliance with UNESCO's Man and Biosphere (MAB) Programme and the International Society for Mangrove Ecosystems (ISME) to undertake this effort. It was clear that a multidisciplinary team of researchers, scientists, and other professionals should evaluate the existing situation and describe a framework for prescriptive remedies; this was realized in the form of an international workshop in March 2000, jointly organized by the three partners. Very appropriately, the islands of Okinawa Prefecture in Japan (where the only mangrove forests in Japan can be found) were selected for this purpose. The collection of papers presented in this book is extremely enlightening in the breadth of the issues covered. The book also raises some issues that warrant closer inspection and, perhaps, thorough scientific investigation. These issues run beyond the traditional studies performed for mangrove ecosystems and offer the opportunity to explore new frontiers.

First and foremost among these is the *carbon-sequestration capacity* of mangroves – both above the surface and below; some preliminary investigations reported in this book show promising results, on the basis of which, more in-depth studies on carbon-sequestration processes and capacity should be undertaken. Findings from such research can have a significant impact on mangrove-restoration activities, particularly if decision makers in the climate-change "regime" can be fully convinced of the potential benefits. As an example, from remains of mangroves under carbon-trading schemes, a new financial source can be tapped. Other social and economic benefits can be gained through such restoration activities, including the provision of employment to communities involved in restoration activities.

The second most interesting area of scientific research for mangroves that emerged during some of the discussions in the workshop is related to recent *advances in biotechnology*. Mangrove plants show a remarkable adaptability to a range of water-salinity levels, and current biotechnological techniques have made it possible to isolate the genes responsible for this salt tolerance. Preliminary research at the Swaminathan Foundation in India has shown that these genes can be successfully transferred to rice species. If such techniques can be fully developed and replicated on a large scale, mangroves may, indeed, prove to be the engine for a second revolution in agricultural production. Needless to say, such technologies are urgently needed in arid and semi-arid regions, where brackish water is often the only option available.

A third area for research is the complete *evaluation of the societal value of mangroves*. This should include the reduction in vulnerability offered by mangroves to coastal communities through protection against extreme climatic events such as cyclones and floods. The "bioprospecting" value of the mangroves – such as utilization of mangrove species for medicinal and pharmaceutical purposes – also remains relatively unexplored. Through extensive research, these benefits need to be evaluated explicitly and tied in to other existing studies for socio-economic valuation of mangrove ecosystems.

In summary, it is quite obvious that mangroves the world over are probably at the highest level of threat, as a direct consequence of a number of human activities in the coastal and adjacent areas. A new vision and appreciation of mangroves is needed to reverse this situation. This appreciation must extend to all levels of society – from small coastal communities, to megacities, to governmental policy makers. International agencies must also realize their potential role in enlightening people on mangroves issues, reversing destructive trends, and protecting the existing precious ecosystems.

Mangrove ecosystems have a great deal to offer us, provided that they are adequately protected and utilized in a sustainable manner. It is hoped that this book will help to trigger effective research and policy development in that direction.

Zafar Adeel
Tokyo, Japan
March 2001

Introduction and keynote presentation

Introduction

Marta Vannucci

There is much knowledge about mangrove ecosystems, but only a small portion of this knowledge is published in scientific journals as an outcome of scientific research. Often the people who have the data are too busy working in the field to write papers. Others who would publish research results have not had sufficiently long experience in the field nor a sufficiently long series of data. Long-term studies on mangrove natural forests and mangrove plantations (as, for instance, in the Sundarbans of Bengal and Matang in Malaysia) show that 10 to 15 or more years of observation and measurements are required, before reliable and well-informed guidelines for sustainable management of the forest can be drawn. Management plans must be revised periodically. Techniques and methodologies cannot always be transferred directly to other places where environmental conditions – including socio-economic interests and pressures – would be different, without previous studies of the new location. On the other hand there is an urgent need to systematize research results for the rational use and management of the various mangrove ecosystems of the equatorial and tropical coastal belts of the world. Hence the importance and opportunity of the workshop organized jointly by United Nations University (UNU), the International Society for Mangrove Ecosystems (ISME) and MAB, the Man and Biosphere programme of the United Nations Educational, Scientific, and Cultural Organization (UNESCO), with the logistic support of the Okinawa Prefectural Government and the University of the Ryukyus.

It is well known that waters from different watersheds carry different types and amounts of inorganic and organic particulate matter. Brackish waters of estuaries, river deltas, and coastal lagoons are further enriched by mixture with marine coastal waters brought in by tidal fluxes. The high productivity of the coastal zone – and, in particular, of the intertidal mangrove forests – is to a great extent due to the nature of brackish waters, where primary and secondary production reach high levels of productivity. The mangrove flora and fauna in turn enrich the mangrove ecosystems through intense recycling. The intertidal belt, periodically covered and uncovered by tides, is particularly vulnerable to violent episodic events such as floods, cyclones, hurricanes, tidal bores, tsunamis, and geomorphological changes of many types, especially at low latitudes. Once the mangrove cover is destroyed, the forest does not easily regenerate spontaneously because environmental conditions change, following the impact of natural and man-made events.

Exchange of matter and energy between the mangrove ecosystem and other ecosystems is intense and uninterrupted, given the high temperature and high light intensity at low latitudes, where the mangrove forest grows to its maximum potential. Changes in remote systems, such as snow-melt from high mountain systems, barrages upstream across feeding rivers, oil spills in the coastal area, or reduction of nutrients due to forest felling, can drastically change the structure and function of the mangrove ecosystem and can positively (or, more often, negatively) affect adjoining systems. The yield and species composition of brackish waters and coastal fisheries are usually the most affected by the degradation of mangrove forests.

Empirical knowledge of the properties of the species of the flora and fauna of the mangroves has come down to us from ancient times. Wherever they occur, the forest and the waters were used by Man and, in places, were managed to a certain extent. Modern scientific research on mangroves started with Linnaeus, who named several species and genera. Research on the biology of the species of the flora and fauna and on the ecology of the ecosystem grew in the nineteenth and twentieth centuries through the establishment of herbaria, nurseries, plantations, and intensive aquaculture in the second half of the twentieth century. Research on the biology of brackish-water organisms started early in the seventeenth and was intensified in the eighteenth century, partly for practical purposes (to control fouling organisms growing on ship hulls and other structures, because they slowed down the sailing ships of the age of discoveries and sea travels). At present, time-series data from field work on natural and plantation mangroves and nurseries are available in a few countries, mainly Bangladesh, India, and Malaysia, but also in Indonesia, Pakistan, Thailand, the Philippines, Viet Nam, and other countries. Usually, the main topics of research are the physiology of plants and animals, due to their intrinsic scientific interest and for practical management purposes. Long-term studies of nurseries and of natural and plantation mangrove forests for cultivation purposes are still scarce, scattered, and unpublished, although they are of fundamental importance when and where sustainable management is the goal. Capture fisheries have long been studied for practical purposes, such as the habits of anadromous, catadromous, and resident species. Captive fisheries at family or village level has long been practised, especially in Asia. Intensive aquaculture developed late, during the second half of the twentieth century, and was pushed to extremes for quick economic gains, causing socio-economic disasters in many places. The scarcity and late start of methodical studies for practical purposes is one reason for the widespread failure of the aquaculture of shrimp ponds dug in former mangrove land and managed for intensive production. Basic scientific research and long-term applied research do not receive the attention they deserve, and important projects may fail to reach their goals because of the lack of sustained funding.

Intensive aquaculture, mainly shrimp farming, has mushroomed since the middle of the last century; the countless failures and few success stories provide useful lessons about what should and should not be done. Several papers discussed at the workshop illustrate this point.

The special complexity of the mangrove ecosystems has drawn the attention of many specialists from different fields of knowledge. This is one reason why so much is known about mangroves, but so little of this knowledge has been knit together into a comprehensive understanding of how the system works. The paucity of interaction among researchers of different disciplines, and the specificity of problems at different places, are some of the reasons why, so far, there is not a lot of wisdom for defining management practices and for developing a workable legislation regarding mangroves of the world. Exercises such as the UNU/ISME Workshop are a serious step towards developing what could be called a "mangrove science" or "mangrove doctrine" for the rational use and management of mangrove ecosystems.

It is high time to bring together scientists and other professionals interested in mangrove studies from different points of view. Taxonomists, physiologists, soil scientists, hydrologists, biochemists, palynologists, microbiologists, geneticists, plant pathologists, foresters, sociologists, and others speak different languages – and so do the coastal dwellers who depend entirely or partially on mangrove sources for their livelihood. It is urgent (but not too late) to harmonize scattered knowledge into a universally understood and meaningful discipline. At present, the conviction is gaining strength that competent scientists from many disciplines can make useful contributions to the conservation of the mangroves of the world. Witness to this, the different activities that have taken place in the near past and present. The title of the workshop that led to the present publication highlights the need of research for conservation. National and international training courses have become more and more specific by focusing on defined subjects, including socio-economic studies and community participation. A three-day symposium was held in 2001, at the University of Tokyo, which covered many different aspects of basic research. In the month of March 2001, the launching of the website of project GLOMIS (Global Mangrove Database and Information System), which is an ITTO/ISME project meant to build an information system particularly useful for poor countries, took place at Okinawa, Japan. Also in 2001, a UNESCO/MAB project called ASPACO (Asia Pacific Co-operation for Sustainable Use of Renewable Resources in Biosphere Reserve and Similar Managed Land) was launched, beginning with an international training course focused on biodiversity, at the Centre for Advanced Studies at Parangipettai, Annamalai University, India. These, and others, are steps in the right direction, but they will need to be correlated efficiently and their scope enlarged. Many competent groups are working in near isolation, linked to each other mainly, or only, by the publication of research papers in international journals. However, international journals very often do not reach the libraries of laboratories and universities of developing countries, where they would be most needed.

Community participation has finally been recognized as the basic and indispensable requirement for dealing meaningfully and efficiently with coastal management, particularly in tropical mangrove ecosystems.

Both forests and fisheries have been grossly overexploited in many parts of the world: the forest (which is the resource basis) is shrinking, quantitative assessments are few, and estimates vary widely. Exercises such as the present workshop and others tend to put matters in their right perspective and call for better estimates and precise assessments. The usefulness of the UNU workshop lies in that it highlighted the gaps and shortcomings of the present trend of taking hurried decisions in all areas of environmental conservation, use, and management. The workshop also pointed to some ways and means to solve problems arising. The post-World-War hurry for "development," or better, "development at all costs," has taken a very heavy toll of the mangroves. Fortunately, there is now a drive towards the restoration of degraded mangrove areas and abandoned shrimp-ponds, but the eagerness to do good work should not lead to blind actions. Someone talking to me once told me "I believe in action," but the question is how to decide on the most appropriate action? Mangroves vary widely from one place to another and sweeping generalizations may lead to gross errors in practice. They also lead to inaccurate estimates of the production potential and productivity of world mangroves.

Mangroves are precious, as Jared Bosire, a trainee from Kenya of the Japan International Cooperation Agency (JICA) mangrove-management training courses said: "Mangroves, these beautiful forests that grow where others can't."

The present volume is divided arbitrarily in three parts, although most papers contain material related to more than one part. The division follows the simple reasoning that the structure of a system should be known in order to understand the function of its elementary parts and of the system as a whole. This knowledge should, in turn, lead to sensible management for defining rational, regional, and worldwide appropriate policies for sustainable production and for realistic enforceable legislation, for scientific research, and for applied site-specific practical results. Ultimately, not only conservation and sustainable utilization of mangroves should be sought but, above all, a healthier and better lifestyle for the coastal dwellers themselves.

Keynote presentation: What we can do for mangroves

Shigeyuki Baba

Introduction

Mangroves are a highly productive ecosystem: the invertebrate and vertebrate fauna of mangrove forests is rich in both number of species and number of individuals of each species. The waters where mangroves grow provide suitable breeding and nursery areas for a large number of fish, shrimps, crayfish, clams, and other aquatic organisms. The same waters are used as feeding, breeding, and resting places for many species of birds, amphibians, reptiles, and terrestrial and aquatic mammals.

Degradation and destruction of mangrove forests is now rampant throughout the tropics and subtropics. Although humans have traditionally used both the direct and indirect benefits offered by the mangrove ecosystem, humans throughout the world often sacrifice long-term hidden benefits and capital wealth for the sake of immediate monetary returns. The consequence of this attitude is that humans will suffer in the future, if we do not act soon and in unison, to protect and restore mangrove forests.

Japan consists of 47 prefectures. Okinawa Prefecture is the southernmost and it is a two-and-a-half-hour flight from Tokyo to Naha, the capital of Okinawa Prefecture, where mangroves are found at present. In Naha, the average mean annual temperature is 22.40°C and the annual rainfall is about 2,036 mm (Okinawa Branch of the Japan Meteorological Association 2000).

The inhabitants of mainland Japan generally do not know much about mangroves because mangroves cannot grow there, except at some places along the southernmost coastline. On the other hand, those who live in Okinawa are quite familiar with mangroves, locally called *hirugi*. Nevertheless, even present-day Okinawans often think that hirugi are merely small, tree-like shrubs that are not very useful. Inhabitants of cities such as Naha sometimes consider hirugi to be useless or even harmful: they assert that hirugi may cause flooding during heavy rains because of dense growth in small rivers. It is sometimes claimed that mangroves have a foul smell, but that is because people illegally dump their garbage where the mangroves grow; the blame should rest on the garbage dumpers rather than on the hirugi, which are equally harmed by such garbage. It is often thought that, without the hirugi, garbage and soil sediments would not be trapped but would be swept out to sea and out of sight and there would not be any offensive smell; however, older inhabitants of Iriomote, Ishigaki, and other remote islands of Okinawa Prefecture, remember that mangroves have been very useful in the past: the bark of hirugi was used for dyeing and providing materials for fishing nets; the timber was used for roofing and other purposes; and mangroves are well known to provide protection against strong winds and waves. Mangroves were highly praised by the ancestors of the older residents.

The word "mangrove" has been used to refer either to the constituent plants of tropical intertidal forest communities or to the community itself (Tomlinson 1986). However, I always recommend that the general public, especially schoolchildren (who are not scientists), think of the word "mangrove" simply as describing plants distributed in the intertidal zone in tropical and subtropical regions. In this way, unnecessary confusion about the meaning of the word mangrove can be avoided.

I would like to emphasize here, once again, the importance of mangroves and the need for people to learn what can be done for mangroves. Clearly, something must be done, because mangrove ecosystems around the world have suffered the consequences of mismanagement or downright destruction, and have been sacrificed for the benefit of various human activities.

The importance of mangroves

In Okinawa, the bark of hirugi has been used for dyeing materials for clothing, for fishing nets, and for the sails of small fishing boats. The timber has been used as roofing material for local houses, for poles in paddy fields, and for other purposes. The bark of hirugi is still extracted from the forest without causing permanent damage to the trees and is used as a natural dye for traditional textiles. The resource is managed in a sustainable manner.

Photos[1] K.1 and K.2 show small mangrove trees at the mouth of the Kokuba and Noha rivers of Okinawa Island. Inhabitants of the island wrongly think that hirugi trees are very small also in the tropics.

Photos K.3 and K.4 show mangroves on Iriomote Island. The total area of mangrove forests in Japan is only 400–600 ha, about 200 ha of which (the largest continuous mangrove area in Japan) is along the Nakama River (photo K.4) on Iriomote. Here, even in the mangrove forest, trees usually do not exceed 10–15 m in height, although they may be several decades old. The specimen of *Heritiera littoralis* shown in photo K.5 could be as old as 400 years and has become both a landmark and a tourist attraction owing to its size and spectacular buttress roots. Equally tall trees of the same species, but much younger, are found next to it, in the same grove.

Whenever I show photo K.6, a photograph taken in Ecuador on the Pacific coast of South America, Okinawans are very surprised. The tree height in Ecuador is generally more than 40 m, and the tallest mangrove tree, measured near the place where this photograph was taken, is about 62 m in height. The tree in photo K.7 is also very tall; this photograph illustrates the true size of the tree because the white speck next to the tree is not a piece of dust but a child in white clothing. These two photographs prove how mangroves can produce valuable timber. With regard to the destruction and degradation of mangroves, photo K.8 shows particularly well how an entire mangrove system has been destroyed for the construction of a luxury resort hotel.

Photo K.9 was taken in Bali, Indonesia. What is shown here looks like paddy-rice fields raised after mangroves were removed by clear-cutting; in fact, these are abandoned shrimp-ponds at a site where the Japan International Cooperation Agency (JICA) implemented a mangrove-restoration project that lasted from December 1992 to November 1999 (Inoue et al. 1999; see also chap. 18 of this volume). The destruction of mangroves for shrimp-pond construction continues not only in Asia but also in Africa and Latin America (photo K.10).

Mangroves are used to provide fodder for goats, camels, buffaloes, and other cattle. Especially appreciated by the animals are the leaves and young twigs of *Avicennia marina* (photos K.11, K.12).

Photo K.13 shows one of the amazing features of mangrove ecosystems: as previously mentioned, the fauna of mangrove forests is rich in species and in the number of individuals of each species. Additionally, the waters around mangroves provide suitable breeding and nursery areas for a large number of fish, shrimp, crayfish, and other aquatic organisms, as well as feeding, breeding, and resting places for many other vertebrate and invertebrate species. It should always be borne in mind that mangroves are a very special ecosystem (photos K.13–K.16).

Regarding forest resources, mangrove forests produce material suitable for thatching (photos K.17–K.19), for firewood (photo K.20), for charcoal production (photos K.21 and K.22), for dyes (photos K.23, K.24), for scaffolding poles and other construction purposes (photo K.25), and for many other uses. Mangrove forests also provide materials for traditional medicines (Aksornkoae 1987).

Another important benefit provided by mangroves is that their forests play a significant and sustainable role in trapping sediments; preventing coastal erosion; and dispersing the energy of tidal storms, typhoons, and strong winds.

I would like to emphasize one more issue, and that has to do with the carbon stock, both above and below ground. We have recently realized that the amount of carbon below ground is much greater than we had expected, which gives an added dimension to the role of mangroves as carbon sinks (Fujimoto et al. 1996, 1999a, b; see also chap. 7, this volume).

What mangroves are, and what we can do for mangroves

I will explain what mangroves are quite simply, since this is not new information. Some scientists are of the opinion that mangroves include more than 100 plant species (Tomlinson 1986), whereas others maintain that there are only 60–70 exclusive mangrove plant species. However, we should always keep in mind that each and every mangrove species is unique and has special characteristics for adaptation to its specific environment. As mentioned above, mangrove forests not merely are forests but also provide the conditions for the creation and establishment of a very special ecosystem – the mangrove ecosystem. Therefore, I enquire, "What can we do for mangroves?" The goals envisaged are depicted in figure K.1: first, we must conserve existing mangrove forests; second, we must utilize mangrove resources in a sustainable manner, and manage them rationally; third, we must rehabilitate mangroves and restore damaged or totally destroyed mangrove forests and their resources.

Photo K.26 shows abandoned shrimp-ponds and photos K.27–K.36 show mangrove nurseries and planting activities. Once mangroves have been planted correctly at suitable sites for each species, they grow well, as shown in photos K.37–K.39.

What is needed to accomplish the three goals identified in figure K.1 brings us to the specific actions indicated in figure K.2. Action 1 is to promote scientific research at universities and research institutes followed by cooperation among scientists, other professionals, and coastal dwellers

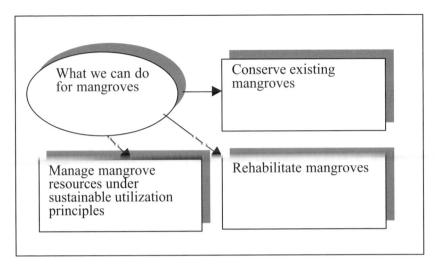

Figure K.1 What we can do for mangroves: goals

through United Nations University, UNESCO, other international and national organizations, and non-governmental organizations (NGOs), through joint projects, such as this workshop and others. We must not only expand our research but also exchange information about what we are doing, so that we have a better understanding of each other and our respective work. Action 2 is information exchange to avoid unnecessary duplication of research and studies, because we must spend time and

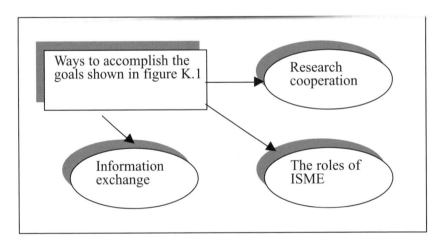

Figure K.2 Means of accomplishing the goals outlined in figure K.1 (ISME, International Society for Mangrove Ecosystems)

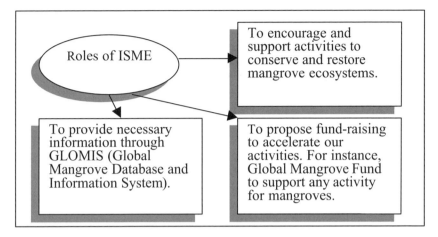

Figure K.3 Roles of the International Society for Mangrove Ecosystems (ISME)

money effectively and efficiently. In my responsibility as Voluntary Executive Secretary of the International Society for Mangrove Ecosystems (ISME), I strongly emphasize Action 3, which is the role of ISME as an international NGO and non-profitable organization (NPO) for the conservation, rational use, and management of mangrove ecosystems.

The roles of ISME are shown in figure K.3: these are to encourage and support activities to conserve and restore mangrove ecosystems; to provide necessary information through the Global Mangrove Database and Information System (GLOMIS); to offer training and education wherever and whenever possible; and to propose fund-raising to accelerate our activities. The main server of GLOMIS is in Okinawa, and GLOMIS has four regional centres located in Brazil, Fiji, Ghana, and India. These four regional centres and the main server exchange necessary information between each other; furthermore, to accelerate our activities, GLOMIS disseminates information to anyone who seeks such information as an end-user. One of the roles of ISME is to promote the conservation and restoration of mangrove ecosystems by all possible means. For this activity we need funds because, whenever we try to undertake new activities, we are always limited by budgetary considerations.

What can people do? They can support our efforts to educate and inform others about the importance of mangroves. They can oppose projects and plans put forward by individuals, corporations, or governments that could jeopardize the inhabitants of mangroves; they can support legislation that protects mangroves and that serves to deter people whose actions threaten mangrove ecosystems; they can also conduct research into valuable use of mangroves and of their resources.

I urge Okinawans to consider that this prefecture consists of a group of small islands and that, although everyone in Okinawa bears the ocean in mind, they do not consider sea-level rise. The truth is that, despite the beauty of our beaches and coastline, if the sea level were to rise about 30 cm, 95 per cent of our beaches would disappear; if the mean sea level were to rise about 65 cm, 100 per cent of our beautiful beaches would disappear. Yet, despite these facts, Okinawans do not pay attention to sea-level rise as one of the consequences of global warming. That is why I present here my proposal for an "Okinawa Mangrove Fund" or "Okinawa Global Environment Fund." If any leader of the Group of Eight (G8) could propose this, it would be most gratifying, not only for those who live with mangroves but also for mangrove researchers, and would ultimately benefit mangroves and the environment worldwide.

I always quote an old Okinawan saying which, translated into English, runs "No forest on the land – no fish in the sea." In other words, "No mangroves along the coast – no fish in the sea." Photo K.40 shows small, low islands facing the threat of sea-level rise in the same way as the islands of Okinawa. If we could plant mangroves around the coastline of small, low islands, we might slow down seawater inundation of the inland area and lessen coastal erosion.

Now it is our turn to conserve and restore mangrove ecosystems, for the sake not only of the mangrove ecosystems themselves but also for us humans, who depend more than we realize on those particular ecosystems.

Conclusions

Nowadays, many mangrove-restoration activities are under way in many countries, but funds are limited and informed local persons are often too few to point decision makers effectively in the right direction. It is important to emphasize that more cooperation and willpower are needed for the exchange of scientific and technical knowledge and know-how and for educational and training activities (such as, for instance, this workshop) to avoid unnecessary duplication of effort, and the waste of time and of limited funds to conserve and restore mangrove ecosystems for future generations.

Note

1. Throughout this volume, photographs have been placed at the end of the chapter to which they refer.

REFERENCES AND FURTHER READING

Aksornkoae, A. 1987. *Mangrove Vegetation and its Utilization. Proceedings of UNESCO Regional Seminar on the Chemistry of Mangrove Plants*. Bangkok, Thailand: Chulalongkorn University, 1–9.

Fujimoto, K., T. Miyagi, T. Kikuchi, and T. Kawana. 1996. "Mangrove habitat formation and response to Holocene sea-level changes on Kosrae Island, Micronesia." *Mangroves and Salt Marshes* 1: 47–57.

Fujimoto, K., A. Imaya, R. Tabuchi, S. Kuramoto, H. Utsugi, and T. Murofushi. 1999a. "Below-ground carbon storage of Micronesian mangrove forests." *Ecological Research* 14: 409–413.

Fujimoto, K., T. Miyagi, T. Murofushi, Y. Mochida, M. Umitsu, H. Adachi, and P. Pramojanee. 1999b. "Mangrove habitat dynamics and Holocene sea-level changes in the southwestern coast of Thailand." *Tropics* 8: 239–255.

Inoue, Y., H.O. Hadiyati, H.M.A. Affendi, K.R. Sudarma, and I.N. Budiana. 1999. *Sustainable Management Models for Mangrove Forests*. Denpasar, Indonesia: The Development of Sustainable Mangrove Forest Management Project, the Ministry of Forest and Estate Crops in Indonesia, and Japan International Cooperation Agency, 214 pp.

Okinawa Branch of Japan Meteorological Association (ed.). 2000. *Okinawa-no-kishoreki [Meteorological Data in Okinawa]* (in Japanese). Okinawa: Okinawa Branch of Japan Meteorological Association, 120 pp.

Tomlinson, B.P. 1986. *The Botany of Mangroves*. London: Cambridge University Press, 413 pp.

Photo K.1 Small mangrove trees (*Kandelia candel*) in Naha City, Okinawa, Japan

Photo K.2 Small mangrove trees in Naha City, Okinawa

Photo K.3 Mangroves on Iriomote Island, Okinawa

Photo K.4 About 200 ha of mangrove forests along the Nakama River, Iriomote Island, Okinawa

Photo K.5 Buttress roots of a very old specimen of *Heritiera littoralis*, Iriomote Island, Okinawa

Photo K.6 *Rhizophora* sp. with very large stilt roots, Esmeraldas, Ecuador

Photo K.7 A child in white clothing stands near the trunk of a very large tree (*Sonneratia* sp.), Indonesia (Photo Izumo)

Photo K.8 Destruction of mangrove forests during construction of a resort hotel (Photo Kogo)

Photo K.9 Abandoned shrimp ponds, Bali Island, Indonesia

Photo K.10 Destruction of mangrove forests during construction of shrimp ponds, Ecuador

Photo K.11 Leaves and young branches of *Avicennia* sp. are an important source of cattle fodder in Iran (Photo Kogo)

Photo K.12 Water buffaloes in Indonesia prefer to eat the leaves and young twigs of *Avicennia marina*

Photo K.13 Fishes gather under mangrove trees on Iriomote Island, Okinawa (Photo Yokotsuka)

Photo K.14 Birds gather on mangrove trees in Florida, USA (Photo Kogo)

Photo K.15 Monkeys in mangroves in Viet Nam; many more were present, although not visible here

Photo K.16 Sea lions taking an afternoon nap in the shade of *Avicennia* trees on the Galapagos Islands, Ecuador

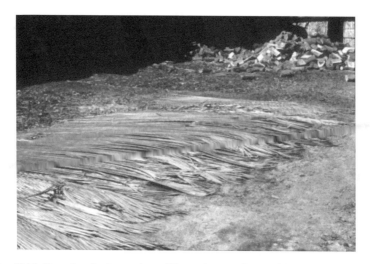

Photo K.17 Fronds of nipa palms (*Nypa fruticans*) provide good materials for thatching in Viet Nam

Photo K.18 Thatching with nipa palm leaves, Viet Nam

Photo K.19 Artistic thatching with fronds of nipa palm, Viet Nam

Photo K.20 Mangrove timber being used as firewood in Fiji

Photo K.21 Bark being peeled off wooden billets that will be burned in kilns to make charcoal, Thailand

Photo K.22 Charcoal kiln, Thailand

Photo K.23 Dyes are extracted from the bark of *Bruguiera gymnorrhiza* and *Rhizophora stylosa* on Iriomote Island, Okinawa

Photo K.24 Dying cloth with extracts from the bark of *Bruguiera gymnorrhiza* and *Rhizophora stylosa*, Iriomote Island, Okinawa

Photo K.25 Poles of *Bruguiera* and *Rhizophora* spp. are used in construction work

Photo K.26 Abandoned shrimp-ponds on Bali Island, Indonesia

Photo K.27 Potted seedlings of *Avicennia* sp., Bali

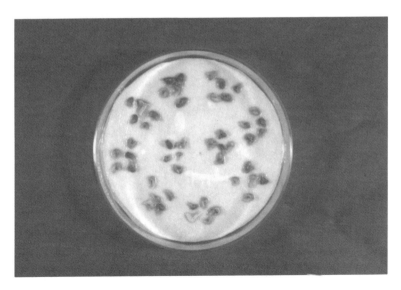

Photo K.28 Seeds of *Sonneratia alba*

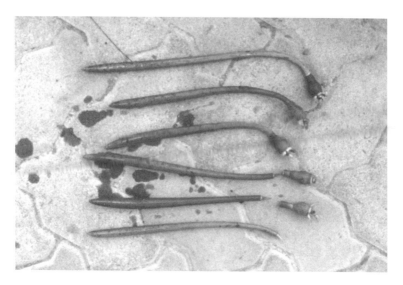

Photo K.29 Propagules of *Rhizophora apiculata*, Bali

Photo K.30 Propagules of *Rhizophora mucronata*, Bali

Photo K.31 Mangrove nursery, Bali

Photo K.32 Experimental mangrove plantation in an abandoned shrimp pond, Bali

Photo K.33 Dense mangrove plantation that will require thinning, Indonesia (Photo: K. Izumo)

Photo K.34 Japanese ladies planting mangroves on Lombok Island, Indonesia

KEYNOTE PRESENTATION 33

Photo K.35 Local inhabitants and Japanese volunteers planting mangroves on Fiji

Photo K.36 Mangrove plantation on the occasion of the Fourth General Assembly of the International Society for Mangrove Ecosystems (ISME), September 1999, Bali, Indonesia

Photo K.37 Mangrove plantation about 20 years old, Viet Nam

Photo K.38 Mangrove plantation about 15 years old, Viet Nam

Photo K.39 Mangrove plantation about 15 years old, Thailand

Photo K.40 Pohnape Island in Micronesia, like other oceanic islands, faces the danger of a rise in mean sea level (Photo Nakamura)

Part I

The mangrove ecosystem: Structure and function

1

Some ecological aspects of the morphology of pneumatophores of *Sonneratia alba* and *Avicennia officinalis*

T. Nakamura, R. Minagawa, and S. Havanond

Introduction

Species of *Sonneratia* and *Avicennia* are usually distributed on the seaward mud-flats of mangrove forests, where the water is shallow.

The normal pneumatophores of these species are slender and cone shaped, standing erect and aligned on the cable roots which spread horizontally in every direction in the soil. These roots may be 1–20 m or more in length, depending on the size and age of the tree. Large pneumatophores may exceed 60 cm in height: the largest that I have seen (a pneumatophore of *Sonneratia caseolaris* in Sumatra, Indonesia) was over 1 m tall.

The shapes of pneumatophores and the patterns of growth and spread of the cable roots vary with the species, age, and habitat of the tree. Environmental factors such as soil characteristics, depth of tidal flood, and tidal regime appear to be the factors with the greatest influence on the growth and shape of this very special root system. Massive waves, sediment deposition on the soil, the nature of the sediments, and other environmental changes may damage and deform the pneumatophores during their growth. The amount and extent of deformation of the pneumatophores may therefore indicate the dynamics at the seaward area of the mangrove community.

The research described here, which focused on the ecology of pneumatophores, was conducted from February to March 1999 at Ranong and at Phang-nga Bay in Thailand. This chapter is based on measurements taken during short visits to Thailand of a month or so.

Methods and results

Age, cable roots, and pneumatophores of Sonneratia alba

It is difficult to determine accurately the age of the individual trees surveyed. Age classes were roughly categorized as follows: (1) large trees with trunk diameter at breast height (DBH) > 30 cm and tree height > 15 m; (2) medium-sized trees with DBH 10–30 cm and tree height 5–15 m; (3) small (young) trees with DBH < 10 cm and tree height < 5 m (it is assumed here that large trees are old and small trees are young).

Cable roots of young trees running parallel to the shoreline are short; they are somewhat longer for medium-sized trees. However, this tendency is not clear for cable roots running seawards or landwards. The cable roots of large trees are obviously long. Although the small sample of trees studied does not permit generalization, it was usually found that the length of cable roots of exceptionally large trees (200 cm DBH) and those of medium-sized trees (13 cm DBH) was almost the same. It is not evident that the cable roots grow longer and extend further as the trees become older, because appropriate individuals between 25 and 40 cm DBH were not found in the study area. It appears that the extent of cable roots of *Sonneratia* species is more related to soil structure and tidal regime than to the age of the tree.

Shape of pneumatophores of Sonneratia alba *and the environment*

The number of pneumatophores is, on average, 2.32/m and 6.07/m in large and young trees, respectively. To study the morphology of pneumatophores the following steps were taken. First, solitary trees were selected for easy identification of their root system. Then, three ropes were stretched in the following directions: parallel to the shoreline; seawards; and landwards. Finally, the shapes of the pneumatophores on the cable roots in each direction were noted.

Figure 1.1 shows the numbers of pneumatophores standing on each one of the three cable roots selected in the three different directions and at various distances from the tree. Pneumatophores with a pointed apex (types A, B, and C) are common, but where the main cable roots possess pneumatophores with a thick, bent, dichotomous or trichotomous, and swollen apex (type D) they appear to have been damaged and deformed in the growth process. In large trees, normal pneumatophores (types A and B) are scattered up to the ends of cable roots, and abnormal types

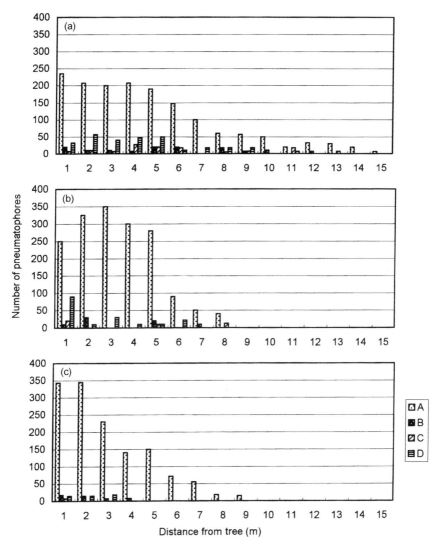

Figure 1.1 Numbers and types (A–D) of pneumatophores at various distances from trees of *Sonneratia alba*

C and D become more numerous. Conversely, young trees have type A pneumatophores evenly distributed on the cable roots; type C and D pneumatophores are extremely scarce. Figure 1.2 depicts the proportions of the various types found on the roots of (a) large, (b) medium-sized, and (c) small trees. This shows that pneumatophores of type A (considered to grow normally) are extremely abundant in the young tree class,

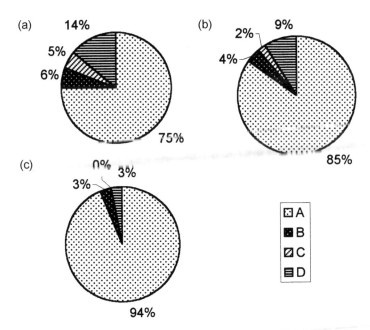

Figure 1.2 Proportions of various types (A–D) of pneumatophores found on the roots of (a) large, (b) medium-sized, and (c) small (young) trees of *Sonneratia alba*

whereas types C and D (which seem to be deformed by external stresses) are frequently found in large trees. This may be a natural phenomenon, because large, old trees are prone to be influenced and deformed by long-term external stresses.

Figure 1.3 shows the results of a comparison of the morphological differences of pneumatophores from Ranong and from Phang-nga, the actual study sites being Hatsai Khao in Ranong and the Phangee River in Phang-nga. The results suggest that mangroves receive greater external stresses in Ranong (which is visited by many tourists) than in Phang-nga. It is also likely that the pneumatophores of *Sonneratia* species in Ranong have suffered additional, major damage as a result of local people gathering shells, oysters, and other products of the mangrove forest, thus adding to the damage caused by boat traffic.

Pneumatophores of *Avicennia alba* and *A. officinalis*

The length of cable roots, the number and size of pneumatophores, and other characteristics of young *Avicennia alba* and *A. officinalis* at Hatsai

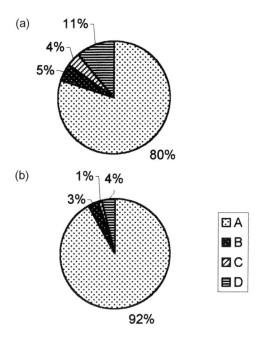

Figure 1.3 Comparison of proportions of pneumatophores of types A–D on roots of *Sonneratia alba* from (a) Ranong and (b) Phang-nga

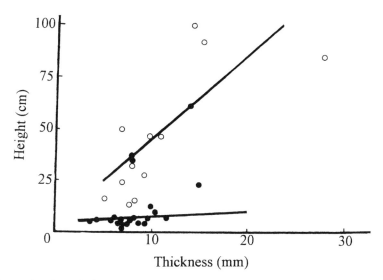

Figure 1.4 Correlation between young tree height and cable-root thickness in (•) *Avicennia officinalis* ($y = 6.4697 \pm 4.1323x$; $r = 0.7483$) and (∘) *A. alba* ($y = 1.8812 \pm 1.28099x$; $r = 0.8160$)

Khao at Ranong were studied for comparison with the pneumatophores of *Sonneratia alba*. Figure 1.4 shows the correlation between young tree height and cable-root thickness. The latter, in *A. officinalis*, varies independently of the height of the tree, which suggests that growth conditions are important factors for this species.

Discussion and conclusions

Of the five species of the genus *Sonneratia* that occur in South-East Asia, *S. alba* and *S. caseolaris* are the most common. *S. alba* grows seawards in the mangrove forest, whereas *S. caseolaris* is more frequently found in the low-salinity waters of the upper part of rivers and creeks. Given its preference for the open front of the mangrove forest, *S. alba* receives greater and more direct impact from waves and winds and from various human activities, including tourism. This study was carried out to investigate the extent to which environmental influences affect the shape of *S. alba* pneumatophores.

On the basis of the results obtained, it is concluded that medium-sized *S. alba* trees produce pneumatophores 25–35 cm long, 4–6 cm thick at the base, with a regular conical shape. There are differences in size and shape, according to growth period and location. Pneumatophores of *S. caseolaris* may become very large, as is shown by those from Delta Upang, Palembag, Sumatra (Indonesia), which may be 1.2 m tall and 12 cm thick at the base; however such growth is exceptional.

2
Introduction of *Sonneratia* species to Guangdong Province, China

Zhongyi Chen, Ruijiang Wang, and Zebin Miao

General information about mangroves in Guangdong Province

Guangdong Province has a long coastline and rich mangrove wetlands. Mangroves, as the main type of intertidal vegetation, were widely distributed along the tidal areas of the province before the 1950s, extending to about 42,000 km^2; now, only one-third – about 14,000 km^2 – is left because of the overutilization of coastline resources and severe destruction of the vegetation.

About 32 species are known in Guangdong Province; these belong to 19 families and 28 genera (Gao 1985; Wang et al. 1998; Wu 1994) (table 2.1). Of these, nine are trees, four are shrubs, and nineteen are herbs. It can be concluded that the biodiversity of the mangroves in Guangdong is relatively restricted. The tree species, which represent 28.1 per cent of the total, are the most effective in protecting the dams and crops and in affecting the local habitat.

Present serious situation

In the 1950s, a project was implemented to create land from the sea by landfill in the coastal zone. For this purpose, local people destroyed

Table 2.1 Mangrove species of Guangdong Province

Division	Family	No.	Species	Character[a]
Pteridophyta	Acrostichaceae	1	*Acrostichum aureum* L.	H
		2	*A. speciosum* Willd.	H
	Blechnaceae	3	*Stenochlaena palustris* (Burm. f.) Bedd.	H
Spermatophyta	Azioaceae	4	*Sesuvium portulacastrum* Linn.	H
	Chenopodiaceae	5	*Arthrocnemum indicum* (Willd.) Moq.	H
		6	*Suaeda australis* (R. Br.) Moq.	H
	Acanthaceae	7	*Acanthus ilicifolius* L.	H
		8	*A. ebracteatus* Vahl	H
	Compositae	9	*Pluchea indica* (Linn.) Less.	H
		10	*P. pteropoda* Hemsl.	H
	Convolvulaceae	11	*Stictocardia tiliaefolia* (Desr.) Hall. f.	H
	Cyperaceae	12	*Cyperus malaccensis* Lam. var. *brevifoliu* Boock.	H
	Gramineae	13	*Paspalum distichum* Linn.	H
		14	*Sporobolus virginicus* (L.) Kunth	H
		15	*Zoysia matrella* (L.) Merr.	H
	Euphorbiaceae	16	*Excoecaria agallocha* Linn.	S
	Goodeniaceae	17	*Scaevola sericea* Vahl.	H
	Malvaceae	18	*Hibiscus tiliaceus* L.	T
		19	*Thespesia populnea* (Linn.) Sol. ex Corr.	T
	Myrsinaceae	20	*Aegiceras corniculatum* (L.) Blanco	H
	Pandanaceae	21	*Pandanus tectorius* Sol.	T
	Papilionaceae	22	*Pongamia pinnata* (L.) Merr.	T
		23	*Derris trifoliata* Lour.	H
	Rhizophoraceae	24	*Bruguiera gymnorrhiza* (L.) Poir.	T
		25	*Ceriops tagal* (Perr.) C. B. Rob.	S
		26	*Kandelia candel* (L.) Druce	S
		27	*Rhizophora stylosa* Griff.	T
	Sonneratiaceae	28	*Sonneratia apetala* Ham.-Buch.	T
		29	*S. caseolaris* (L.) Eng.	T
	Sterculiaceae	30	*Heritiera littoralis* Dryand.	T
	Verbenaceae	31	*Avicennia marina* (Forsk.) Vierh	S
		32	*Clerodendron inerme* (L.) Gaertn.	H

a. H, herb; S, shrub; T, tree.

extensive mangrove areas along the seashore. The tidal areas were expected to become productive agricultural land, but nothing could be grown on them except seaweed. In the 1980s more and more shrimp- and fish-ponds were constructed to obtain yet more sea products. This caused further deterioration of the coastal environment, and few mangroves were left, unless they were protected in time. Mangroves now occupy less than 14,000 km² in Guangdong Province; natural disasters have been more frequent and violent than previously; and both the life and property of local inhabitants are again at risk from typhoons.

Reforestation

Because of the serious situation that was created, a mangrove-restoration programme was started in 1993. *Sonneratia apetala* was selected as the most suitable species for sylviculture because it is cold resistant and fast growing. It was therefore planted in the northern districts of the province, and time has shown that it can grow even further north than expected.

Procedure

Collecting seeds of Sonneratia apetala

About 20 seedlings were taken from Hainan Island and were later planted in Leizhou, a city near the sea, in the autumn of 1993. At the same time, about 50 kg of ripe *Sonneratia* fruits were collected from the Hainan mangrove nature reserve.

Preservation of seeds

To ensure a high germination rate, the fruits were preserved in fresh water, which was changed regularly; the seeds were thus kept in a healthy state for about two months. In springtime (during the warmer weather) the seeds were washed out of the fruits and plots were prepared for sowing.

Selection of site for sowing

The plots for sowing should be on the inner beach, away from the sea and above the high-water mark, to prevent the seeds being washed away. A freshwater well should be available nearby to water the plots. The seeds germinate better if they are watered with fresh water or water with salinity of less than 5 parts per thousand (ppt).

Preparing the basic medium in the seed-plots

Soft, fertilized, and sterilized soil is the best medium to ensure germination in the seed-plots. Before sowing, a thin layer of such previously prepared soil must be spread in the seed-plots; a thin layer of soil should cover the seeds after sowing.

Nursery arrangements

The seeds must be spread evenly in the seed-plots and sprayed with fresh water when the beds dry out. A metal grid should be placed over the seed-beds to keep away crabs and rodents. During cold weather, the beds should be kept warm. When the seedlings grow up to 70 cm tall they can be planted in estuaries or tidal land along the seashore. When the first six or eight leaves have formed, the seedlings must be transplanted into plastic bags of nutrient soil, one seedling per bag.

Planting

Selecting the planting site

Estuaries with deep, muddy soil, rich in humus and with the correct salt content, are appropriate for the seedlings. If the salt content is higher than 15 ppt, the seedlings will not survive or thrive because they cannot adapt to the new environment quickly on transference from the seed-plot.

Seedling protection

The newly planted seedlings should be protected in their first two years, for they may be destroyed by local people or tidal action. Furthermore, marine animals such as crabs and barnacles may nibble or adhere to the stems of the seedlings, thus preventing the seedlings' growth or even killing them; if this situation is very serious, a pesticide with low toxicity can be used. Another adverse factor is the winter cold: the longer the low temperature persists, the more seedlings will die.

Growth of Sonneratia apetala

The growth rate of *S. apetala* is very high: from table 2.2 it can be seen that the mean height of 2-year-old trees is 3.5 m in Furongwan and Jiepao, but 3.5-year-old trees in Leizhou were 11.5 m tall. Moreover, the highest growth rate is from the second year after planting to late in the third year (fig. 2.1). After four or five years' rapid growth, *S. apetala* grows slowly and flourishes. More branches appear and more aerial roots emerge from the ground.

Table 2.2 Height (m) of *Sonneratia apetala* after planting

Plot	Date:	1995 Sep	1996 Apr	1996 Oct	1997 May	1997 Nov	1998 Jul	1999 May
A	Leizhou	0.9	1.3	2.2	4.0	6.5	8.0	11.5
B	Furongwan				0.8	1.0	2.5	3.5
C	Jiepao				0.8	1.4	2.1	3.5

Development of the plantation

In 1993, Gaoqiao mangrove nature reserve was selected as our first planting place; however, the seedlings all died because of the high salt content and serious damage by barnacles in 1994. Subsequent selection of appropriate planting places was more careful; thus, good progress was made in developing planting techniques (Chen Zhongyi et al. 1999). Now, about 80 km² man-made *S. apetala* forest belts have been established along the coastline of the Zhanjiang District.

Effect of mangrove introduction

Vegetation

After the introduction of *S. apetala*, the local mangrove species harboured a more diverse range of species. Before introduction, only some *Kandelia* or *Avicennia* species were present.

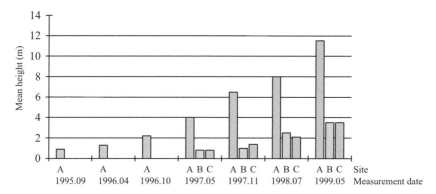

Figure 2.1 Mean growth height of *Sonneratia apetala* at (A) Leizhou, (B) Furongwan, and (C) Jiepao at various times after planting

Animals

More animals, such as fish and crabs, are now more common in these areas. The fallen leaves and fruits provide rich food for neritic creatures, and many birds nest in the new habitat.

Dam protection

The fast-growing trees consolidate the dams, and the cost of dam reinforcement is thereby reduced. In 1995, a typhoon (no. 9507) passed through the Zhanjiang District, destroying many towns and dams, as well as the eucalyptus trees; however, dams near the *Sonneratia* forest belt suffered little damage.

REFERENCES

Chen, Z. et al. 1999. *Sonneratia* reafforestation – Restoration of mangroves in China. Ecotone VIII. "Enhancing coastal ecosystem restoration for the 21st century". Seminar Abstracts, 23.
Gao, Y. 1985. "Mangroves in Guangdong." *Tropical Geography* 5(1): 1–8.
Wang, R. et al. 1998. "Chromosome counts on Chinese mangroves." *Journal of Tropical and Subtropical Botany* 6(1): 40–46.
Wu, T. 1994. *A Checklist of Flowering Plants of Islands and Reefs of Hainan and Guangdong Province*. Science Press.

3

Research into, and conservation of, mangrove ecosystems in Indonesia

Aprilani Soegiarto

Introduction

The mangrove ecosystem in Indonesia originally covered an area of 4,250,000 ha and was distributed over all major islands of the Indonesian archipelago. Unfortunately, owing to various human-induced pressures, probably now only 2,500,000 ha are standing.

For centuries, the Indonesian people have utilized mangroves for firewood, charcoal, tannin dyes, timber, boat construction, and many other purposes. Traditional uses of mangrove resources currently continue side by side with large-scale and intensive exploitation using high-capital investment and technologies, such as charcoal production and raw material for pulp for paper-mills and chipboard. In recent years, efforts to convert mangrove land for other uses, such as fish- or shrimp-ponds or industrial and human settlements, have been increasing in number and size. Most environmental problems in mangrove ecosystems are related to intensive exploitation or conversion of mangrove land to other uses. In order to retain the ecological function of the mangrove ecosystem, the Indonesian government has designed several ways of protecting and preserving mangroves – for example, by establishing coastal green belts, mangrove-conservation areas, national parks, reforestation of degraded or denuded areas, and the development of the *tambak-parit* system of land use. The role of research for the conservation of mangrove ecosystems in Indonesia has not been neglected.

The Indonesian archipelago is located along the equator between two continents – Asia and Australia – and two oceans – the Pacific and the Indian. It is the largest archipelago in the world, comprising more than 17,000 islands. About 70 per cent of the archipelago is covered by sea. The coastline of the archipelago, which is more than 81,000 km long, is probably one of the longest national coastlines in the world.

It has been noted that the marine and coastal region of South-East Asia, where the Indonesian archipelago is situated, is probably one of the world's most productive areas. Blessed with a warm, humid, equatorial climate and high rainfall, the region enables coral reefs and mangrove ecosystems to flourish. Because of the economic benefits that could be derived from these rich and diverse ecosystems, the coastal zones of South-East Asia are densely populated: over 70 per cent of the population of the region lives in coastal areas, resulting in a high level of exploitation of natural resources. Population pressure associated with several economic activities has caused large-scale destruction and serious degradation of the coastal and marine environment; increasing pollution in the last decade, both on land and at sea, has greatly compounded such problems.

Mangrove ecosystems grow most extensively in the humid tropics, along the protected coastal shores with muddy to sandy bottoms. Ecologically, mangroves represent a rather sharp transitional gradient between the marine and the freshwater environments; therefore, only those flora and fauna with broad physiological limits of tolerance can survive in such an environment. In Indonesia, the mangrove ecosystem develops well along the leeward coastlines of most of the large islands.

The following is a brief review of the present knowledge of the mangrove ecosystem in Indonesia, including the thrust of its research programme and the efforts made to conserve this important resource. For more detailed reviews see Kartawinata et al. (1979), Soegiarto (1980), Soemodihardjo (1984), and Soemodihardjo et al. (1992).

Brief description of the mangrove ecosystem

Extent of coverage and distribution

Earlier authors, including MacNae (1974) and Van Steenis (1941), estimated that the mangrove ecosystems in Indonesia covered an area of only 1–2 million hectares; however, intensive surveys and mapping made in recent years, using satellite imaging and aerial photography, have produced an official figure of 4,250,000 ha (Darsidi 1984). Mangroves are distributed throughout the islands of the Indonesian archipelago. Unfortunately, however, owing to increasing human-induced pressures, their area has been steadily decreasing. At present over 50 per cent of the mangrove ecosystem is recorded in Irian Jaya and is still in pristine condition; in contrast, the mangrove ecosystem most disturbed is in Java, where only about 50,000 ha of mangroves are left. The largest remaining mangrove stand in Java is found in Segara Anakan and Cilacap, on the south coast of Central Java. In other parts of Indonesia the mangroves have been moderately utilized (Noor 1995; Soegiarto 1991).

Biodiversity

The mangrove ecosystems harbour highly diverse species, Indonesian mangroves having the highest degree of biodiversity. Kartawinata et al. (1979) and Giesen (1993) have recorded 189 plant species – including 80 species of trees, 24 species of lianas, 41 species of ground-covering plants, 41 species of epiphytes, and 3 species of parasites.

The mangrove ecosystem also serves as a habitat for birds, insects, mammals, reptiles, molluscs, and crustaceans. Although the existence of an exclusive mangrove fauna is probably rare, some species are very closely associated with mangrove communities. Organisms flourishing in brackish and marine conditions are the principal mangrove fauna, molluscs, crustaceans, and fishes being the dominant groups. The terrestrial elements include proboscis monkeys, wild pigs, mouse deer, wild cats, and rats. A variety of reptiles also live in the mangrove ecosystem, such as monitor lizards, snakes, and estuarine crocodiles. A more complete list of the mangrove fauna in Indonesia can be found in Kartawinata et al. (1979).

Function, utilization, and conversion

The mangrove ecosystem has many functions, such as coastal stabilization, prevention of erosion, and provision of suitable conditions for spawning. It is a nursery for many species of economic importance and probably also serves as a pollution trap. In addition, mangroves have many industrial uses, such as the production of firewood, of charcoal, and of logs and raw materials for the manufacture of paper and chipboard.

For centuries the Indonesian people have utilized mangroves for firewood, charcoal, tanning dyes, timber, and boat construction. Genera frequently used for these purposes include *Rhizophora*, *Bruguiera*, *Ceriops*, *Avicennia*, *Nypa*, and *Oncosperma*. The leaves of *Nypa* spp. (common name nipa) are used for various purposes – such as for thatched roofs, baskets, and cigarette papers. The stalks of the inflorescence are cut and the sap is tapped for making brown sugar or the fermented palm wine *arak*.

The Riau province in Sumatra has long been a centre for charcoal production from mangrove wood. The product is exported to Singapore, Malaysia, and Hong Kong. Export of mangrove logs is also a lucrative business: the logs originate from Sumatra, Sulawesi, and Kalimantan; they are exported to Japan and Taiwan for conversion into wood chips. A Japanese–Indonesian joint enterprise for the production of wood chips has begun operations in Indonesia; another example is Chipdeco, in East Kalimantan; an additional paper-mill has been established in Irian Jaya by a joint Indonesian–American company.

In general, mangrove soils are marginal for agriculture; nevertheless, conversion of mangrove land to agriculture has occurred in some parts of Indonesia. Examples can be found in Cilacap, Indramayu, and Sukabumi, all of which are in Java. Another agricultural use of mangrove land is for coconut plantations, such as those in Riau Province and in other parts of Sumatra, where they are developed by coastal communities of the Bugis people.

Extensive mangrove-land conversion has occurred for brackish-water fish-ponds or *tambak*. There are about 200,000 ha of tambak, distributed mainly in Java, Sulawesi, and Sumatra. Traditionally, the tambak ponds are used to raise milkfish (*Chanos chanos* Forsk.), whereas *Mugil cephalus*, *Tillapia mossambica*, and penaeid shrimps (*Penaeus* spp.) are secondary products. In recent years, in order to raise foreign-exchange earnings from the fisheries sector, Indonesia has constructed more than 100,000 ha of new tambak for rearing the highly priced tiger prawn (*Penaeus monodon*). Although much of the new tambak is located in the outer islands, a certain percentage is still constructed in Java, such as on the north coast of West Java and in East Java.

In recent years the demand for mangrove land for human settlements, industrial estates, ports, and other facilities has increased rapidly. Around the cities of Jakarta and Surabaya, a large portion of mangrove land has been (or is being) converted into housing complexes, industrial sites, warehouse compounds, harbour developments, and recreational areas (Soemodihardjo 1984).

Research programmes

Since the nineteenth century, the mangrove ecosystem has been of interest to biologists and naturalists. In recent years, research activities have also been carried out on the ecology, flora and fauna associations, population dynamics, and interaction with other coastal ecosystems, such as coral reefs, sea-grass beds, and estuaries.

In order to generate interest and coordinate research activities on mangrove ecosystems in Indonesia, a National Mangrove Committee was established in 1978. At present, the committee has some twenty members, representing government agencies, research institutions, and the universities that have research and development programmes pertaining to mangrove ecosystems.

The following are some of the committee's objectives (Soegiarto 1984):
- To coordinate research programmes on mangroves.
- To prepare lists of institutions and agencies dealing with mangrove programmes.
- To prepare a directory of mangrove scientists.
- To compile and update a bibliography on mangrove research published in Indonesia.
- To organize every four years a scientific seminar on mangrove ecosystems. The purpose of this seminar is to review the state of knowledge, to evaluate research results, and to plan and provide directions for future research programmes. The first seminar was held in February 1978 in Jakarta; the sixth seminar was held in Pekan Baru, Riau Province, Sumatra in September 1998. Proceedings of the seminar are published within a year of the seminar itself.
- To grant financial support for research within limits.
- To develop and enhance regional and international cooperation on the mangrove ecosystem. Through this function the Indonesian National Mangrove Committee cooperates, interacts, and also acts as the national contact with various regional and international organizations, such as the International Society for Mangrove Ecosystems (ISME), the United Nations Educational, Scientific, and Cultural Organization (UNESCO), and the Association of South-East Asian Nations (ASEAN).

On the basis of a series of meetings and seminars, the Indonesian National Mangrove Committee has established a general framework of research that can be used by government agencies, research institutions, and universities in developing their research activities. These broad research topics include coastal land-use mapping; a resources inventory, including studies of biotic characteristics and processes; resources management; and silvicultural research and conservation. Other studies include topics on environmental degradation and pollution; socio-economic studies; mangrove utilization; and the formal and non-formal education of, and enhancement of environmental awareness by, coastal communities at large.

Also identified were studies on the importance of microclimate; groundwater level and salt-water intrusion in the mangrove forest stands; and research on the ecological effectiveness of the green belt, both from the economic point of view and as a physical protection of the shoreline.

Socio-economic and cultural studies on the communities that live in and around the mangrove ecosystem are in need of further research. Lastly, studies on the role of the mangrove ecosystem in relation to health problems, such as malaria, are also encouraged.

Management and conservation

Responsibility for managing mangrove forests in Indonesia is with the Department of Forestry and Crop Estates. In Java, this function is delegated to the "Perum Perhutani" (State Forest Corporation) and to the "Perum Inhutani" for mangrove areas outside Java. One of the positive management activities of the "Perum Perhutani" is the reforestation of damaged mangrove forests in Java. Almost 3,000 ha of damaged mangrove forest have been rehabilitated in Cilacap on the south coast of Java, in the last three years. Over 10,000 ha of disturbed mangrove areas on the north coast of West Java have also been replanted. A new approach called *tambak empang parit* (channel fish-ponds) has been initiated on the north coast of West Java to take socio-economic factors into account. This new approach is a means by which people who live in the surrounding areas may catch or culture fish in the broad channels around the rehabilitated mangrove forest areas for their daily livelihood (Soegiarto 1984).

Rhizophora and *Avicennia* seedlings were used in the rehabilitation programme for the north coast of West Java, whereas *Rhizophora* and *Bruguiera* spp. were used in the Cilacap mangrove area. The survival rate of replanting was reported to be between 60 and 75 per cent.

Encouraged by the success of these rehabilitation programmes, a number of coastal villages in various parts of Indonesia have started their own replanting programmes, under the guidance and assistance of local universities, governments, and non-governmental organizations.

In order to preserve the mangrove ecosystem, the Indonesian government – through the departments of Agriculture and of Forestry – has regulated the felling of mangrove forests. According to these regulations, a 50–200 m green belt of mangrove must be retained along the coast and a 10–20 m green belt along river banks. This green belt serves not only to preserve the ecological functioning of the mangrove ecosystem but also to ensure the natural regeneration process of the mangroves in the surrounding area.

Another effort in conserving the mangrove ecosystem in Indonesia is by establishing protected and conservation areas. To ensure that the natural ecosystem will still exist for the use and benefit of future generations, a number of nature reserves have been established in Indonesia. Many of these reserves are located in the coastal area and some have mangrove components. There are 13 reserves which have the protection of mangroves as their principal aim (Soegiarto et al. 1982). Mangroves are also found bordering 14 other protected areas where they are of secondary interest. Further, nine proposals for mangrove reserves have been approved by provincial governments and more proposals are currently being processed by the Directorate-General of Forest Protection and Nature Conservation. Surveys and studies on proposed sites are being carried out jointly by the Directorate of Nature Conservation, the Centre for Research and Development in Oceanology, the Center for Research and Development in Biology, some universities, and the World Wildlife Fund of Indonesia.

Conclusions

Originally, Indonesia possessed about 4,250,000 ha of mangrove ecosystems, distributed along the inner coasts of most of the major islands. Some of these have been overutilized and are in a critical state; some are moderately utilized; and the rest are still in pristine condition. Unfortunately, owing to increasing environmental and economic-development pressures, the mangrove areas have steadily declined.

The responsibility of managing the mangrove ecosystem in general is in the hands of the Department of Forestry and Crop Estates. However, research and development activities are carried out by many government agencies, research institutions, and universities. In order to coordinate the research programmes and to develop and enhance cooperation among the institutions dealing with mangrove ecosystems at national, regional, or international levels, a National Mangrove Committee has been functional since 1978.

A number of programmes are under way to protect and conserve the mangrove ecosystem – such as promoting rehabilitation and replanting, establishing *tambak empang parit* schemes, establishing coastal green belts, establishing protected and conservation areas, and developing community-based management.

REFERENCES

Darsidi, A. 1984. "Mangrove forest management in Indonesia." In: S. Soemodihardjo, P. Wiroatmodjo, S. Bandijono, M. Sudomo, and Suhardjono (eds). *Proceedings of Seminar on the Mangrove Ecosystem II, Baturaden, 1982.* Jakarta: LIPI, 19–26.

Giesen, W. 1993. *Indonesia's Mangroves: An Update on Remaining Area and Main Management Issues.* Bogor: Asia Wetlands Bureau.

Kartawinata, K., S. Adisumarto, S. Soemodihardjo, and I.G.M. Tantra. 1979. "Status pengetahuan hutan bakau Indonesia (The state of knowledge on the Indonesian mangrove forest)". In: S. Soemodihardjo, P. Wiroatmodjo, S. Bandijono, M. Sudomo, and Suhardjono (eds). *Proceedings of the Seminar on the Mangrove Ecosystem, Jakarta, 1978.* Jakarta: LIPI, 21–39.

MacNae, W. 1974. *Mangrove Forests and Fisheries.* FAO/IOFC/DEV/74/73. Rome: FAO, 35 pp.

Noor, Yus Rusila. 1995. "Mangrove Indonesia, Pelabuhan bagi keanekaragaman hayati: Evaluasi keberadaannya pada saat ini (Indonesian mangroves, harbour for biodiversity: An evaluation on their current existence)". In: S. Soemodihardjo, P. Wiroatmodjo, S. Bandijono, M. Sudomo, and Suhardjono (eds) *Proceedings of the Fifth Seminar on the Mangrove Ecosystem, 1995.* Jakarta: LIPI, 299–309.

Soegiarto, Aprilani. 1980. *Status Report of Research and Monitoring of the Impact of Pollution on Mangrove and its Productivity in Indonesia.* UNEP/FAO Regional Southeast Asian Consultative Meeting on Mangrove Ecosystem, Manila, 4–9 February 1980. Manila: FAO. 51 pp. + appendix.

Soegiarto, Aprilani. 1984. "State of knowledge and research programmes on mangrove ecosystem in Indonesia." Paper presented at the Meeting on Asean–Australian Cooperative Programs on Marine Sciences. Project 2: Living Resources in Coastal Areas with Emphasis on Mangrove and Coral Reef Ecosystems, Bangkok, Thailand, 25–26 June 1984. 13 pp. (unpublished).

Soegiarto, Aprilani. 1991. "Research and management of mangrove ecosystem in Indonesia." In: Sanjay U. Deshmukh and Rajeshwari Mahalingam (eds). *Proceedings of the Foundation Workshop for Establishing a Global Network of Mangrove Genetic Resources Centers for Adaptation to Sea Level Rise, Madras, India. Proc. No. 2.* Madras, India: CRSARD, 43–52.

Soegiarto, Aprilani, Soewito, and R.V. Salm. 1982. "Development of marine conservation in Indonesia." Invited, unpublished paper. Session IV B (Indonesia), World National Parks Congress, Bali, Indonesia, October 1982, 16 pp.

Soemodihardjo, S. 1984. "Mangrove information system in Indonesia." Country Report, presented in the Asian Regional Workshop on Mangrove Information, Manila, 30 April–2 May 1984, unpublished.

Soemodihardjo, S., P. Wiroatmodjo, A. Abdullah, I.G.M. Tantra, and A. Soegiarto. 1992. "Condition, socio-economic values and environmental significance of mangrove areas in Indonesia." In: B.F. Clough (ed.). *The Economic and Environmental Values of Mangrove Forests and their Present State of Conservation in the South-East/Pacific Region. Mangrove Ecosystems Technical Reports, ITTO-ISME TS-12 Vol. 1,* 17–40.

Van Steenis, C.G.G.J. 1941. "Kustaanwas en mangrove." *Naturwissenschaften Tijdschrift Nederlands Indie* 101: 82–83.

4

Status of Indian mangroves: Pollution status of the Pichavaram mangrove area, south-east coast of India

AN. Subramanian

Introduction

The Indo-Malaysian area is considered to be the evolutionary cradle of the mangrove ecosystem (Krishnamurthy 1993), because it is widely believed that mangrove plants evolved first in this area and later spread to other regions of the tropics. At present the Indo-Pacific region is known for its luxuriant mangroves: mangrove forests are most abundant in South-East Asia, in particular. The Sunderbans regions of India and Bangladesh together form the single largest block of mangroves worldwide.

India has a very long coastline with very variable ecological features. A recent Government of India publication states that the total length of the Indian coastal zone, including the island territories, is 7,516.6 km. To undertake national programmes for the conservation of Indian mangroves and their associated flora and fauna, there is an urgent need to highlight or emphasize the significance of Indian mangroves. For such a project, the Indian mangroves should be treated as critical ecosystems and studied in detail, case by case.

History of Indian mangroves and research

Indian mangroves have a long history. From as early as the Upper Cretaceous to the Miocene period, mangrove vegetation was prolific along the Indian coast, as is evident from the available fossil specimens.

59

During more recent times, the Portuguese were among the first Europeans to visit and settle in the Indian coastal area, towards the end of the fifteenth century. They learned how to use mangroves to create rice–fish–mangrove farms and taught this technique to the people of such African countries as Angola and Mozambique (Vannucci 1997). The first scientific report on Indian mangroves – *Hortus Bengalensis* – was published in 1814 by Roxburgh, who described therein the flora of mangroves of the Sunderbans. In 1891, Schimper published a classical work, "*Die Indo-Malayischen Strandflora*," which was followed by publications by Clarke (1869), Prain (1903), Blatter (1905), Cooke (1908), and others. The practices of mangrove management were also first started in India in the nineteenth century in the Sunderbans, for timber (Vannucci 1997); these were later adapted to local conditions in Malaysia and Indonesia.

Hooker, in his book *Flora of British India* (1872–1897), explained several geographical divisions of India on the basis of plant types. The surveys of Champion (1936) and also of Gamble (1936) are among studies of the phytogeographic regions of pre-independence India.

One decade after Indian independence, the importance of these areas was recognized. The Government of India has introduced a scheme for conservation and protection of the mangrove ecosystem and set up a National Experts Panel on the mangrove ecosystems. Prior to this, several meetings and symposia were held and the published proceedings of these have highlighted the status of Indian mangroves and their decline and have also warned of the urgent need for conservation and preservation of these endangered coastal ecosystems.

A status report on *Mangroves of India* was brought out in 1987 by the Ministry of Environment and Forests (Anon. 1987), which described the mangrove stands all along the east and west coasts and also in the coastal regions of the Andaman and Nicobar islands, as well as on Minicoy Island in the Lakshadweep Archipelago.

As a result of several activities carried out by the Expert Committee, several major and minor publications have appeared (Bhosale 1986; Deshmukh and Balaji 1994). It was felt necessary to update the status report and many more studies have been published since then (Anon. 1997, 1998, 1999; Gopal and Krishnamurthy 1993; Kannan 1996; Kathiresan 1992, 1995; Untawale 1987). The coordination of global mangrove research through the International Society for Mangrove Ecosystems (ISME) based at Okinawa (Japan) was first presided over by Dr M.S. Swaminathan, an Indian. The M.S. Swaminathan Research Foundation (MSSRF) initiated pioneering research on mangrove genetics in 1991. It has also identified 23 sites from nine countries (Cameroon, India, Indonesia, Malaysia, Papua New Guinea, Pakistan, the Philippines, Senegal,

Table 4.1 Area (km^2) under mangroves in India

No.	State	Area (km^{2a})	Area (km^{2b})
1	West Bengal	4,200	1,619
2	Andaman and Nicobar	1,190	770
3	Orissa	150	187
4	Andhra Pradesh	200	480
5	Tamil Nadu	150	90
6	Gujarat	260	1,166
7	Maharashtra	330	138
8	Goa	200	5
9	Karnataka	60	19
10	Kerala	Sparse	Sparse
Total		6,740	4,474

Source: *a.* Untawale (1987); *b.* Nayak (1993).

and Thailand) for the future isolation, from mangrove species, of genetic material conferring salt tolerance and its transference to crop plants. The four sites identified in India are Bhitarkannika, Coringa, Goa, and Pichavaram.

Mangroves in India

According to a Status Report published by the Government of India (Anon. 1987), the total area of mangroves in India was calculated to be about 6,740 km^2; this comprised about 7 per cent of the world mangroves and covered 8 per cent of the Indian coastline (Untawale 1987). Recent Indian remote-sensing data (Nayak 1993) have shown that the total area of mangroves has decreased to 4,474 km^2 (table 4.1). This decrease in mangrove area may be due to several causes, including:

- grazing by domestic cattle and exploitation of mangrove woods for fuel and timber;
- neo-tectonic shifting of river courses;
- reduction of upstream freshwater discharge because of dam and reservoir construction;
- rapid trend of reclamation of mangrove forests for habitation;
- discharge of pollutants from cities and industrial areas;
- other causes.

Recent data available from the Forest Survey of India, Dehra Dun, shows an extent of 4,827 km^2 as the total mangrove area in India. The statewise distribution of mangroves is depicted in figure 4.1: of the total mangrove area, 57 per cent is on the east coast, 23 per cent on the west

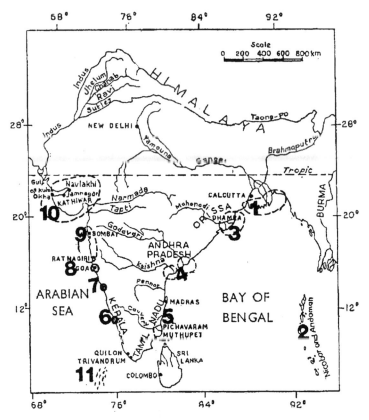

Figure 4.1 Statewise distribution of mangroves in India. Source: Naskar and Mandal (1999)

coast, and the remaining 20 per cent on the Bay of Bengal islands (Andaman and Nicobar).

Mangroves are prominent on the east coast of India because of the nutrient-rich alluvium brought in by the rivers – Ganges, Brahmaputra, Godavari, Mahanadi, Krishna, and Cauvery – and the perennial supply of fresh water to the coastal deltas, as a result of which only deltaic mangroves exist on the shores of the Bay of Bengal. On the other hand, the west coast is characterized by funnel-shaped estuaries of the rivers Tapti and Narmada, with typical creeks and backwaters; hence, backwater–estuarine-type mangroves occur on the coasts of the Arabian Sea (Naskar and Mandal 1999). Gopal and Krishnamurthy (1993) have shown that insular-type mangroves are present in the Andaman and Nicobar islands, where the lagoons and islets support a rich mangrove flora.

The western seaboard

Mangrove vegetation on the western coastline extends over the maritime states of Gujarat, Maharashtra, Karnataka, Goa, and Kerala. According to Sidhu (1963), the total coverage of mangroves on the Indian west coast was about 1,140 km^2 at that time, but the Status Report published by the Government of India (Anon. 1987) emphasized that west-coast Indian mangroves, 24 years after Sidhu's assessment, covered an area of only 850 km^2.

Gujarat State on the west coast has the second-largest area of mangroves along the Rann of Kutch and Kori creek. In Maharastra and Goa, mangroves occur especially in large stands along the Mandovi River estuary, the Kundalika River estuary, and several creeks. Mangroves in Karnataka cover an area of 6,000 ha, and the stretches of mangroves in Kerala State are at present very sparse.

The Andaman and Nicobar islands in the Bay of Bengal harbour a rich variety of mangroves covering about 770 km^2, which are found along the creeks and near bays and lagoons, the dominant species being *Rhizophora mucronata*, *Avicennia* spp., and *Ceriops tagal* (Singh et al. 1986).

The eastern seaboard

Along the east coast of India, the Sundarbans region in West Bengal State has an area of 4,200 km^2 (Anon. 1987), which, together with its area in Bangladesh, forms the largest block of mangroves in the world. In Orissa State, mangroves are present on the Mahanadi delta, with *Rhizophora mucronata*, *Excoecaria agallocha*, and *Avicennia* spp. as the dominant floral species. In Andhra Pradesh State, dense mangrove vegetation is found on the western side of the Krishna River delta. Mangroves in Tamil Nadu State occur on the Cauvery River delta. The Pichavaram area has a well-developed mangrove forest with *Rhizophora* spp., *Avicennia marina*, *Excoecaria agallocha*, *Bruguiera cylindrica*, *Lumnitzera racemosa*, *Ceriops decandra*, and *Aegiceras corniculatum* as the most frequent species. Mangroves can also be found near such places as Vedaranyam, Kodaikarai (Point Calimere), Muthupet, Chatram, and Tuticorin. Although the Pichavaram mangrove area is small, it has been well studied with regard to all aspects of biology, chemistry, and microbiology by scientists from the Centre of Advanced Study in Marine Biology, the Department of Botany and Faculty of Agriculture of Annamalai University, and also, more recently, by the MSSRF, Chennai (formerly Madras). Tissot (1987) and Caratini, Blasco, and Thanikaimoni (1973) used palynological methods to investigate the changes that have taken place in the vegetation of the Kaveri River delta over the last 2,000 years.

Pichavaram mangroves

The Pichavaram mangrove forest is located about 200 km south of Chennai on the south-east coast of India. This forest is actually enclosed between two prominent estuaries – the Vellar River estuary in the north and the Coleroon River estuary in the south. The Vellar Coleroon estuarine complex forms the Killai backwater and Pichavaram mangroves (fig. 4.2).

The Pichavaram mangrove (latitude 11°21′N; longitude 79°50′E) occurs in the lower area of the Vellar–Coleroon estuarine complex. It covers an area of 1,100 ha, with a heterogeneous mixture of mangrove elements. The source of fresh water to this mangrove is from both rivers and from rains, while that of seawater is from the Bay of Bengal.

The whole of the mangrove comprises about 51 small and large islands, ranging in area from 10 m^2 to 2 km^2. The mangrove soil is formed mainly by the alluvium carried by the rivers, enriched with detritus derived mainly from the mangrove plants. About 40 per cent of the total area is covered by waterways, 50 per cent by forest, and the rest by mud-flats and by sandy and salty soils. Numerous creeks, gullies, and canals traverse the mangroves, ranging in depth from 0.5 to 1.5 m. A major irrigation channel discharges agricultural waste water from the upper reaches to this mangrove.

The Pichavaram mangrove did not receive much attention during the pre- and post-independence periods. A map published by the Cuddalore District authorities in 1882 is the document which was first made public. Subsequently, but not until the later part of the 20th century, Thirumalairaj (1959) explored the Pichavaram mangrove and Venkatesan (1966) listed the floral communities of the region in relation to environmental factors.

The French Institute at Pondicherry is one of the pioneering institutes in Pichavaram exploration and has contributed several publications on the wealth of the mangroves (Blasco 1975; Meher Homji 1979). The Centre of Advanced Study in Marine Biology, from the outset of its inception in 1961, has been involved in various research activities relating to Pichavaram mangroves. Water quality, floral and faunal composition, microflora, ichthyofauna, bioactive substances from mangroves, fishery resources, larval development, heavy-metal and organochlorine residues, methanogens, Cyanobacteria, wood biodeterioration, and UV radiation are all studied by this centre. During the 1990s, the MSSRF established

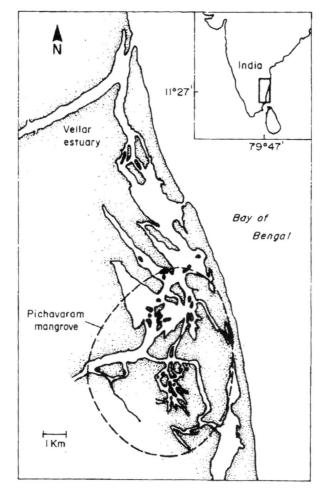

Figure 4.2 Pichavaram mangrove forests on the south-east coast of India

a mangrove Genetic Resource Conservation Centre at Pichavaram by adopting 50 ha of forest area. In addition, the Centre for Water Resources, Anna University, Chennai has produced satellite images through remote-sensing techniques.

The mangrove flora in India comprises 35 species, of which 33 (16 genera and 13 families) occur along the east coast (Kathiresan 1998). The east coast of India, like the Andaman/Nicobar islands, has a high species diversity. Pichavaram has 14 exclusive mangrove species (Kannupandı and Kannan 1998): *Avicennia marina* alone constitutes nearly 30 per cent of the total population, followed by *Bruguiera cylindrica* (17 per cent), and *Avicennia officinalis* (16 per cent). The population density of other species is poor, and many of the species are on the verge of extinction. According to a recent survey, nearly 62.8 per cent of the Pichavaram mangrove forests were degraded between 1897 and 1994. Tissot (1987) investigated the changes that have taken place in the vegetation of the Kaveri delta over a period of 2,000 years and found that the breadth of the sand bar protecting the mangrove areas from wave action was reduced by 550 m between 1970 and 1992.

The substantial reduction in the forest cover is partly due to frequent cyclones, which occur at least every other year: they have devastated several mangrove species and reduced the total area from 4,000 ha at the beginning of the century to nearly 1,100 ha at present. As a result, many plants previously recorded from that mangrove have completely vanished. For example, pollen analysis of the sediments from Pichavaram have shown that *Sonneratia* was abundant here in the past (Caratini, Blasco, and Thanikaimoni 1973), but is now on the verge of extinction. Certain species, such as *Xylocarpus granatum*, *Rhizophora stylosa*, and *Bruguiera gymnorrhiza*, which previously were harvested from this mangrove, are not now present (Kannupandi and Kannan 1998). Furthermore, most of the individuals of *Rhizophora* spp. are aged and their rate of reproduction is also low at Pichavaram. This species seems to be on its way to extinction at this mangrove, being replaced by the much more dynamic (but less valuable) *Avicennia marina* (Kannupandi and Kannan 1998).

From the Pichavaram mangrove ecosystem about 100 species of diatoms, 20 species of dinoflagellates, 40 species of tintinnids, 30 species of copepods, 30 species of prawns, 30 species of crabs, 30 species of molluscs, and 200 species of fish have been recorded (Anon. 1987).

The wealth of fauna

It is estimated that 8 tonnes of organic plant detritus per hectare per year are produced by the withered mangrove leaves in the Pichavaram mangrove. These leaves are colonized by bacteria and fungi, which, in turn, are consumed by protozoans. All of these give rise to rich, particulate, organic matter, forming the source of food for such creatures as crabs, worms, shrimps, and small fishes. These, in turn, are preyed upon by more than 60 species of larger fish. Several species of small organisms live under the prop root system; these form the food for post-larval, juvenile, and adult fishes, and prawns such as *Penaeus indicus*, *P. mono-*

don, *P. semisulcatus*, *Metapenaeus dobsoni* and *M. monoceros*. The mangroves are used as a breeding ground for prawns such as *Macrobrachium* spp. and for certain fishes.

Studies of the food-web pattern and of fish eggs and larvae (Prince-Jeyaseelan 1981, 1998) have shown that 77 species of fishes belonging to 51 genera and 33 families are living in this area. Commercially important fish species belonging to the families Mugilidae, Chanidae, Clupeidae, Pomodasyidae, and Gerridae are harvested from this mangrove. Prince-Jayaseelan (1998) has written a *Manual of Fish Eggs and Larvae from Asian Mangrove Waters*, chiefly based on species from these waters. "Seeds" for aquaculture of fish and prawn species are abundantly available.

Chemical studies

The results are available of many studies on the chemical aspects of this mangrove from the 1960s onwards – that is, from the inception of the Centre of Advanced Study in Marine Biology of Annamalai University. Simultaneous investigations have been carried out on the Vellar–Coleroon estuarine complex, particularly on the four aquatic biotopes – namely, the mangrove waterways, the adjacent Killai Backwater, the Vellar and Coleroon estuaries, and the neritic waters of the Bay of Bengal; these have shown which environmental and meteorological factors exert most influence on this mangrove biotope (Anon. 1987).

With normal (130 cm) and abundant (> 130 cm) rainfall, the mouths of the mangrove and estuaries become deep and remain open to the sea. They receive a considerable volume of marine coastal water inflow, particularly at high tide; hence, the mangrove ecosystem functions rather like a coastal marine ecosystem. However, in years when rainfall is poor, the mangrove ecosystem functions rather like a freshwater or limnetic ecosystem. The marine coastal water influence in the mangroves is felt for a distance of about 10 km; the maximum depth in the waterways is about 1 m; the salinity varies from 0 to 34 ppt; the annual temperature range is from 20° to 34°C; the water pH ranges between 7.60 and 8.50; the dissolved oxygen content is, on average, about 4.5 ml/litre.

With regard to pollution studies, the earlier studies by Ramadhas, Rajendran, and Venugopalan (1975), Sundararaj (1978), Ramadhas (1977), Subramanian (1982), Subramanian and Venugopalan (1983) and Subramanian, Subramanian, and Venugopalan (1981) are worth mentioning.

Subramanian (1982) and Subramanian and Venugopalan (1983) found that the salt-excreting type of mangrove, *Avicennia marina*, accumulates

Table 4.2 Physical and chemical parameters observed in the Pichavaram mangroves during June and October 1999

No.	Parameters	June 1999		October 1999	
		Minimum	Maximum	Minimum	Maximum
Physical parameters					
1	Atmospheric temperature (°C)	31	39.5	29	32
2	Water temperature (°C)	29	35.5	28.5	31
3	Light intensity (lux)	476	4,710	325	985
4	Depth (m)	0.29	64.5	0.3	71.71
5	Extinction coefficient	0.5	4.25	1.97	5.85
Chemical parameters					
6	pH	6.8	8.1	7.04	7.88
7	Dissolved oxygen (ml/litre)	3.0	5.5	2.97	4.68
8	Total dissolved solids (mg/litre)	3.45	8.5	5.2	12.3
9	Salinity (ppt)	22.0	35.0	6.39	29.9
10	Chlorinity (ppt)	12.2	19.39	3.32	16.4
11	BOD (ppm)	2.85	5.2	1.95	4.54
12	POC (mg/litre)	14.6	38.2	3.40	7.63
Nutrients					
13	Nitrite (μg/litre)	3.9	9.9	4.2	12.0
14	Nitrate (μg/litre)	62.7	123	56.9	266
15	Inorganic phosphate (μg/litre)	16.4	55.8	22.9	74.1
16	Total phosphate (μg/litre)	21.1	52.7	38.2	84.5
17	Reactive silicate (μg/litre)	120	176	119	359
18	Calcium (mg/litre)	320	480	160	520
19	Magnesium (mg/litre)	820	1,280	380	896

Trace metals (ppb)					
20	Zinc	9.64	22.2	35.0	56.2
21	Lead	0.48	1.12	1.0	1.76
22	Copper	6.4	9.7	12.2	16.6
23	Cadmium	0.15	0.30	0.52	0.92
24	Iron	212	448	634	820
25	Aluminium	125	350	420	585
Sediment					
1	Temperature (°C)	28	32.5	29	31.0
2	pH	7.4	8.2	6.9	7.6
Nutrients					
3	Total nitrogen	0.31	1.21	0.68	2.15
4	Total phosphate	0.051	0.96	0.085	0.164
5	Total organic carbon	3.85	8.12	1.2	3.5
Trace metals (ppm)					
6	Zinc	18.525	55.5	0.12	62.2
7	Lead	BDL	24.03	1.46	10.4
8	Nickel	7.93	17.68	10.8	37.4
9	Cobalt	2.53	6.63	3.8	26.0
10	Cadmium	BDL	BDL	0.3	13.52
11	Manganese	52.0	252.5	4.98	438.0
12	Copper	6.15	13.12	12.94	85.6
13	Iron	5,300	10,275	8,080	52,000
14	Aluminium	3,125	8,750	8,089	46,100

Source: T. Balasubramanian 1999, pers. com.
BOD, biological oxygen demand; POC, particulate organic carbon; BDL, below detection limit.

more iron and phosphorus in the leaves than such salt-excluding species as *Rhizophora mucronata*; they also found a clear seasonal fluctuation. Subramanian (1982) found a very clear seasonal fluctuation in the concentrations of Fe, Mn, Cu, and Zn in the environmental compartments (such as the dissolved, particulate, and sediment fractions). This author also found that the sediments of mangroves (which are normally rich in organic matter) are a prime sink for these elements, which are brought in by fresh water. Generally, the reactive forms of trace elements in the water were higher during the monsoon season in this mangrove, indicating the importance of fresh water in enriching these waters with essential nutrients.

Creatures such as the oyster *Crassostrea madrassensis* and the polychaete *Nereis costae* retained a higher concentration of trace elements in the monsoon months. Again, in the case of mangrove plants also, salinity seemed to be the dominant factor in controlling the uptake and accumulation of such elements (Subramanian 1982). Recent data obtained from Pichavaram mangroves of several physical and chemical parameters (table 4.2) showed that this mangrove is still pristine and the values are very much within optimal levels (T. Balasubramanian 2000, pers. com.).

Earlier workers, such as Karthikeyan (1988), expressed concern after finding considerable concentrations of such pesticides as DDTs, lindane (γ-HCH) and heptachlor in the environmental and biological samples collected from the Pichavaram mangrove and the adjacent Vellar estuary. Later, the recent works of Babu Rajendran (1994), Babu Rajendran and Subramanian (1997), and Ramesh et al. (1990, 1991) have shown that the concentrations in the Vellar–Coleroon estuarine complex (comprising the Pichavaram mangrove) have not increased during the past decade.

In spite of their increased usage in the past decade, the concentrations of the two major pesticides used until recently in India – DDTs (pp'-dichlorodiphenyltrichloroethane) and HCHs (hexachlorocyclohexane) – in the Pichavaram mangrove waters (fig. 4.3) and sediments (fig. 4.4) are within acceptable limits. There was clear seasonal variation, showing the dispersion of these volatile residues via monsoon rains or atmospheric passage (Takeoka et al. 1991) (fig. 4.5).

Management perspectives for Pichavaram mangroves

Research by scientists of the Centre of Advanced Study in Marine Biology and the MSSRF has shown that contamination by various chemicals is not at abnormal levels and is not adding pressure to this mangrove ecosystem.

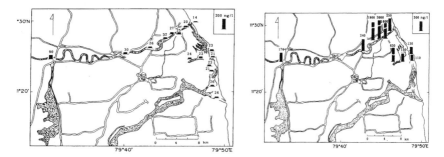

Figure 4.3 Spatial distribution of γ-HCH in the Pichavaram mangrove and Vellar estuary during the dry (left) and wet (right) seasons. Source: Ramesh et al. (1990)

However, despite the economic, ecological, and human values of this mangrove, its areal extent is gradually being reduced because of human interference, cattle grazing, and the illegal felling of trees. Further, reduced freshwater supply to the tail-end area of the Cauvery delta and Vellar River may also be considered as one of the main causes of the degradation of the Pichavaram mangrove ecosystem. An increase in human population around Pichavaram, and aquacultural practices, have also drastically reduced the mangrove vegetation cover. Moreover, recently Pichavaram has become one of the tourism "wonderlands" from all over the country, putting increased pressure on this ecosystem. Comparison of the old Indian topography with the recent satellite imagery of IRS–IB LISS II showed shrinkage of this forest area from 4,000 ha in 1897 to 1,100 ha at present. The setting up of several industries in nearby areas, recently, may put further pressure on this ecosystem.

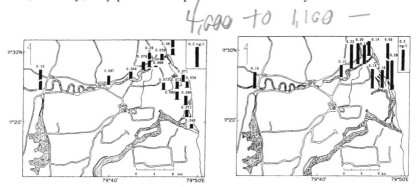

Figure 4.4 Spatial distribution of DDT in the Pichavaram mangrove and Vellar estuary during the dry (left) and wet (right) seasons. Source: Ramesh et al. (1990)

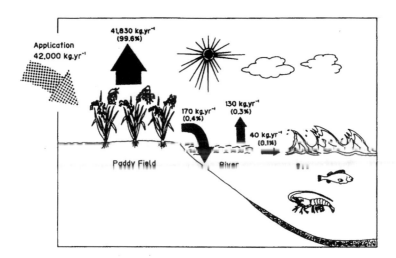

Figure 4.5 Schematic representation of the flux of γ-HCH in the Vellar watershed. Source: Takeoka et al. (1991)

Recently, the Department of Ocean Development of the Government of India has taken up a project jointly with the Centre of Advanced Study in Marine Biology, Annamalai University, in Pichavaram mangrove under the Integrated Coastal Marine Area Management (ICMAM) project for defining ways of effective management of this ecosystem, apart from the efforts of the MSSRF. The efforts of the Centre with the help of a project from Tamil Nadu State Council for Science and Technology have become fruitful in developing the mangroves in the nearby areas of the Vellar estuary (K. Kathiresan, pers. com.). Every possible effort is called for by these institutions to protect the existing remains of Pichavaram mangroves. The practical steps taken by the scientists of both institutions have resulted in the afforestation of the denuded areas and development of the mangroves in the nearby estuarine locations. A recent survey has shown that, in the Pichavaram mangrove region, the forest cover is increasing at a rate of 1 per cent every year (K. Kathiresan 2000, pers. com.).

REFERENCES

Anon. 1987. *Mangroves of India: Status Report.* New Delhi: Government of India, Ministry of Environment and Forests, 150 pp.

Anon. 1997. *Mangroves of India – State of the Art Report (1987–1996).* Annamalai University: ENVIS Special Publication, 61 pp.

Anon. 1998. *An Anthology of Indian Mangroves.* Annamalai University, ENVIS Special Publication, 66 pp.

Anon. 1999. *How to Know the Mangroves? (Annotated Checklist of Indian Mangroves).* Annamalai University, ENVIS Special Publication, 29 pp.

Babu Rajendran, R. 1994. "Baseline studies on DDTs and HCHs in the lower reaches of Kaveri and Coleroon rivers and fishes of Tamil Nadu coast, India." PhD thesis, Annamalai University, India, 123 pp.

Babu Rajendran, R. and AN. Subramanian. 1997. "Pesticide residues in water from river Kaveri, South India." *Chemistry and Ecology* 13: 223–236.

Bhosale, L.J. (ed.). 1986. *The Mangroves. Proceedings of a National Symposium on Biology, Utilisation and Conservation of Mangroves.* Kolhapur, India: Shivajee University.

Blasco, F. 1975. *Mangroves in India. French Institute of Pondicherry. Travel Section, Science and Technology* 14: 1–80.

Blatter, E.J. 1905. "The Mangrove of the Bombay Presidency and its Biology." *Journal of the Bombay Natural History Society* 16: 644–656.

Caratini, C., F. Blasco, and G. Thanikaimoni. 1973. Relations between the pollen spectra and the vegetation of a South Indian mangrove. *Pollen et Spores* 15: 281–292.

Champion, H.A. 1936. "A preliminary survey of the forest types of India and Burma." *Indian Forestry Record (N.S.)* 1: 286.

Clarke, C.B. 1869. "Presidential address to the Linnean Society on the Sunderbans of Bengal." *Proceedings of the Linnean Society of London* 14–29.

Cooke, T. 1908. *The Flora of Presidency of Bombay.* London: Taylor and Francis.

Deshmukh, S. and V. Balaji. (eds). 1994. *Conservation of Mangrove Forest Resources – A Training Manual. ITTO–CRSARD Project.* Madras, India: M.S. Swaminathan Research Foundation, 487 pp.

Gamble, J.S. 1936. *Flora of the Presidency of Madras.* Calcutta: Botanical Survey of India, (Reprinted Edition: 1967). Vol. 3, 493–1389.

Gopal, B. and K. Krishnamurthy. 1993. In: D.F. Whigham, D. Dy Kyjovas, and S. Henjny (eds). *Wetlands of the World.* Netherlands: Kluwer Academic Publishers, 345–414.

Hooker, J.D. 1872–1897. *Flora of British India,* Vols I–VII. Botanical Society of India, Special Publication.

Kannan, L. 1996. *Application of Remote Sensing Techniques to Coastal Wetland Ecology of Tamil Nadu with Special Reference to Mangroves. Final Report of the Ministry of Environment and Forests, Government of India project,* 31 pp.

Kannupandi, T. and R. Kannan. 1998. "Hundred years of Pichavaram mangrove forest." *An Anthology of Indian Mangroves.* Annamalai University, India: ENVIS Special Publication, 66 pp.

Karthikeyan, E. 1988. "Organochlorine pesticides in the tropical estuary (The Vellar), mangrove (Pichavaram) and swamp (Kodikkarai) in the east coast of India." PhD thesis, Annamalai University, India, 151 pp.

Kathiresan, K. 1992. "Foliovory in Pichavaram mangroves." *Environmental Ecology* 10: 988–989.

Kathiresan, K. 1995. "Quo vadis, mangroves?" *Seshaiyana* 3: 25–28.

Kathiresan, K. 1998. *India – A Pioneer in Mangrove Research Too*. Annamalai University, India, ENVIS Special Publication, 66 pp.

Krishnamurthy, K. 1993. "The mangroves." *ENVIS Newsletter* 1: 1–3.

Meher Homji, V.M. 1979. "On the subtropical climate vegetation of the Indian subcontinent." *Geology Today* 8: 137–148.

Muniyandi, K. 1985. "Studies on mangroves of Pichavaram (Southeast coast of India)." PhD thesis, Annamalai University, 215 pp.

Naskar, K. and R. Mandal. 1999. *Ecology and Biodiversity of Indian Mangroves*. Vols I and II. New Delhi, India: Daya Publishing House, 754 pp.

Nayak, S.T. 1993. "Remote sensing applications in the management of wetland ecosystem with special emphasis on management of mangrove ecosystem and coastal protection." Visakhapatnam, India, 21 pp.

Prain, D. 1903. "Flora of the Sunderbans." *Record of Botanical Survey of India* 2: 231–370.

Prince-Jeyaseelan, M.J. 1981. "Studies on ichthyofauna of the mangroves of Pichavaram (India)." PhD thesis, Annamalai University, India, 290 pp.

Prince-Jayaseelan, M.J. 1998. *Manual of Fish Eggs and Larvae from Asian Mangrove Waters*. Paris: UNESCO, 193 pp.

Ramadhas, V. 1977. "Studies on phytoplankton nutrients and some trace elements in Portonovo waters." PhD thesis, Annamalai University, India, 135 pp.

Ramadhas, V., A. Rajendran, and V.K. Venugopalan. 1975. "Studies on trace elements in Pichavaram mangroves (South India)." In: I.G. Walsh, S. Snedaker, and H. Teas (eds). *Proceedings of the International Symposium on the Biological Management of Mangroves*. Gainesville, Florida: University of Florida, 1: 96–114.

Ramesh, A., S. Tanabe, H. Iwata, R. Tatsukawa, AN. Subramanian, D. Mohan, and V.K. Venugopalan. 1990. "Seasonal variation of persistent organochlorine residues in Vellar river waters in Tamil Nadu, South India." *Environmental Pollution* 67: 289–304.

Ramesh, A., S. Tanabe, H. Murase, AN. Subramanian, and R. Tatsukawa. 1991. "Distribution and behaviour of persistent organochlorine insecticides in paddy soil and sediments in the tropical environment: A case study in south India." *Environmental Pollution* 74: 293–307.

Roxburgh, W. 1814. *Hortus Bengalensis or a Catalogue of Plants growing in the Honourable East India Company Botanical Garden at Calcutta*. Serampore Mission Press, 105 pp.

Schimper, A.F.W. 1891. "Die Indo-Malayischen Strandflora." *Botanische Mitteilungen Tropischen* 3: 32–67.

Sidhu, S.S. 1963. "Studies on the mangroves." *Proceedings of the National Academy of Sciences of India* 33: 129–136.

Singh, V.P., L.P. Mall, A. Garge, and S.M. Pathak. 1986. "Some ecological aspects of mangrove forest of Andaman islands." *Journal of the Bombay Natural History Society* 525–537.

Subramanian, AN. 1982. "Some aspects of cycling of iron, manganese, copper, zinc and phosphorus in Pichavaram mangrove." PhD thesis, Annamalai University, India, 252 pp.

Subramanian, AN., B.R. Subramanian, and V.K. Venugopalan. 1981. "Seasonal variation of iron in plankton and mangrove leaves in relation to environmental concentration." *Journal of Environmental Biology* 2: 29–40.

Subramanian, AN. and V.K. Venugopalan. 1983. "Phosphorus and iron distribution in two mangrove species in relation to environment. Mahasagar." *Bulletin of the National Institute of Oceanography* 16: 183–191.

Sundararaj, V. 1978. "Hydrobiological studies in the Vellar–Coleroon estuarine system." PhD thesis, Annamalai University, India, 101 pp.

Takeoka, H., A. Ramesh, H. Iwata, S. Tanabe, AN. Subramanian, D. Mohan, A. Mahendran, and R. Tatsukawa. 1991. "Fate of HCH in the tropical coastal area, South India." *Marine Pollution Bulletin* 22: 290–297.

Thirumalairaj. 1959. "Mangrove forests of Tanjore Division." *Proceedings of the Mangrove Symposium, Calcutta,* 18–19.

Tissot, C. 1987. "Recent evolution of mangrove vegetation in the Cauvery delta: A palynological study." *Journal of the Marine Biological Association of India* 29: 16–22.

Untawale, A.G. 1987. "Conservation in Indian mangroves. A national perspective." In: T.S.S. Rao, R. Natarajan, B.N. Desai, G. Narayanaswamy, and S.R. Bhat (eds). *Special Collection of Papers to Felicitate Dr. S.Z. Qasim on his Sixtieth Birthday,* 85–104.

Vannucci, M. 1997. "Supporting appropriate mangrove management." Intercoast network. Special Edition, 1: 1, 3, 42.

Venkatesan, K.R. 1966. "The mangroves of Madras State." *Indian Forester* 92: 27–34.

5

The role of aquatic animals in mangrove ecosystems

Shigemitsu Shokita

Introduction

The overall productivity of mangrove ecosystems is usually high, as is the level of primary productivity, leading to a rich food web and economically rewarding commercial fisheries.

The ecological roles of mangroves are varied. They are important nursery, breeding, feeding, and spawning grounds for many brackish-water animal species, many of them of great economic importance, such as many bivalves, shrimps, crabs, and fishes. A positive correlation between the inshore fisheries and shellfish catch with the extent of mangrove cover has been demonstrated in Okinawa, South-East Asia, Australia, and elsewhere. In order to preserve and enhance the fishery resources in mangroves and tropical coastal areas in general, it is important to protect and restore, where necessary, the natural environment of the mangroves. The following topics are focused on here: mangal-dwelling taxa; distribution and abundance of macrobenthos; macrofaunal habitats and dominant species; tree-dwelling fauna; stock enhancement of economically important species; feeding habits; and the food chain and the role of crabs in the decomposition process of mangrove litter.

Research aims, objectives, and findings

Mangal-dwelling taxa (photos 5.1–5.4)[1]

The macrobenthos

The intertidal upper zones of the tropical and subtropical sheltered marine shores and coral reefs of the Indo-Pacific Region are occupied by mangrove trees and populated by invertebrates and fishes. The mangal-dwelling macrobenthos consists of Coelenterata, Platyhelminthes, Nemertina, Nematoda, Sipunculoidea, Mollusca, Annelida, Arthropoda, Brachiopoda, Echinodermata, and Hemichordata. Among these, molluscs and crustaceans are the dominant groups (Frith, Tantanasiriwong, and Bhatia 1976; Sasekumar 1974; Shokita et al. 1983) (table 5.1).

The macrobenthos of Okinawan mangrove swamps consists mainly of molluscs and crustaceans, which account for approximately 40 and 55 per cent, respectively, of the entire population. The dominant species of crustaceans are brachyuran crabs, most of which belong to either the Grapsidae or the Ocypodidae families. Grapsid crabs usually inhabit mangrove forests, whereas ocypodids dwell mainly in areas near to the forest edge and mud-flats or along channels. This distribution is similar to that of other mangal swamps in South-East Asia, the Pacific, Australia, and Africa (Jones 1984; MacNae 1968; Sasekumar 1974; Warner 1977). About 90 species of crustaceans have been reported from Okinawan mangal areas (Nakasone 1977; Shokita et al. 1998).

The mangal-dwelling fishes (photos 5.5, 5.6)

About 88 species and 68 genera from 39 families have been identified from the mangals of Okinawa and of the Yaeyama Islands (Ishigaki and Iriomote) (Shokita et al. 1988). The families (and number of species) found in the Okinawan mangals are as follows: Carcharhinidae (1), Orectolobidae (1), Dasyatidae (1), Clupeidae (1), Elopidae (1), Muraenidae (2), Chanidae (1), Plotosidae (1), Synodontidae (1), Hemiramphidae (2), Fistulariidae (1), Sygnathidae (3), Atherinidae (2), Mugilidae (4), Sphyraenidae (1), Ambassidae (1), Serranidae (1), Kuhliidae (3), Apogonidae (1), Carangidae (4), Leiognathidae (3), Gerreidae (4), Mallidae (1), Monodactylidae (1), Lutjanidae (5), Pomadasyidae (3), Teraponidae (1), Sparidae (2), Scatophagidae (1), Chaetodontidae (1), Cichlidae (1), Pomacentridae (2), Acanthuridae (1), Siganidae (3), Gobiidae (19), Platycephalidae (1), Paralichtyidae (1), Bothidae (1) and Tetraodontidae (4). Among these, gobiid fishes are the most abundant, accounting for 19 species.

The faunal categories of the mangal fishes established by Gunderman and Popper (1984) and Por (1984) consist of (1) true resident species, (2)

Table 5.1 Macrofaunal community at the mangal areas of South-East Asia and Okinawa

Taxa	Malaya[a]	Thailand		Okinawa (Iriomote Island)	
		Ranong[b]	Phuket[c]	Funaura[d]	Shiira River[e]
CNIDARIA					
Actiniaria	0	2	4	1	1
	(0)[f]	(1.4)	(2.9)	(1.3)	(1.2)
PLATYHELMINTHES	0	0	1	0	0
	(0)	(0)	(0.7)	(0)	(0)
NEMERTINEA	0	1	6	0	0
	(0)	(0.7)	(4.3)	(0)	(0)
NEMATODA	0	0	0	0	1
	(0)	(0)	(0)	(0)	(1.2)
SIPUNCULIDEA	1	1	3	0	0
	(1.2)	(0.7)	(2.2)	(0)	(0)
MOLLUSCA					
Gastropoda	26	30	24	20	20
Bivalvia	1	25	19	12	11
	(30.6)	(39.3)	(30.9)	(42.1)	(36.5)
ANNELIDA					
Polychaeta	9	27	22	4	3
	(10.6)	(19.3)	(15.8)	(5.3)	(3.6)
ARTHROPODA	48				
Balanomorpha		2	2	2	1
Tanaidacea		0	0	0	0
Isopoda		1	2	1	1
Amphipoda		0	1	1	1
Caridea		10	3	6	9
Anomura		7	7	7	6
Brachyura		29	41	22	31
Stomatopoda		1	1	0	0
	(56.5)	(35.7)	(41.0)	(51.3)	(57.6)
BRACHIOPODA	0	0	1	0	0
	(0)	(0)	(0.7)	(0)	(0)
ECHINODERMATA					
Holothurioidea	0	1	1	0	0
Asteroidea	0	1	0	0	0
Ophiuroidea	0	1	1	0	0
	(0)	(2.2)	(1.5)	(0)	(0)
HEMICHORDATA	0	1	0	0	0
	(0)	(0.7)	(0)	(0)	(0)
Total	85	140	139	76	85

Source: Modified from Shokita et al. (1983).
a: Sasekumar (1974); b: Frith, Tantanasiriwong, and Bhatia (1976); c: Shokita et al. (1983); d: Shokita and Nishijima (1983); e: S. Shokita, unpublished data.
f: Percentages in parentheses.

closely associated species, (3) loosely associated species, and (4) euryhaline species of freshwater origin.

All species of diadromous fish and crustaceans must pass through estuaries during their migration between marine and fresh waters, and most spend a considerable amount of time in the estuarine environment. These species may be (1) anadromous, migrating upstream, (2) catadromous, migrating towards the sea, or (3) amphidromous with alternate behaviour (Day et al. 1987).

Estuaries, including the mangal ecosystem, are utilized by diadromous fish in the following ways (representative fish species from each category of Okinawan mangals are shown in parentheses):
1. Freshwater fishes occasionally enter brackish water (*Oreochromis mossambicus*);
2. Truly estuarine species spend their entire lives in estuaries (*Zenarchopterus dunckeri* and *Periophthalmus vulgaris*);
3. Estuarine–marine species use the estuary primarily as a nursery ground, usually spawning and spending much of their adult life at sea, but often returning seasonally to estuaries (*Siganus guttatus*, *Lutjanus argentimaculatus*, *Carcharinus leucus*, others);
4. Marine species pay regular seasonal visits to estuaries, usually as adults searching for food (*Dasyatis sephen*);
5. Anadromous and catadromous species pass through in transit (*Anguilla japonica* and *A. marmorata*);
6. Occasional visitors appear irregularly and accidentally (*Plotosus lineatus*, *Fistularia pelimba*, others).

Distribution and abundance of macrobenthos

Mangal study sites were established at Smare Kaow and Ranong (fig. 5.1). At Smare Kaow (fig. 5.2), one transect was made across an *Avicennia marina* forest from a shrimp-farm seaward. Five sampling points were established on the transect. The soil at each station was very muddy and fine-grained. At Ranong, nine transects were made across each zone from the mangal forest to the outer mangal muddy tidal flat to examine the zonal and vertical distributions of macrobenthos.

Soil samples were collected near each quadrat to examine the relationship between benthic animals and the substratum. Dried soil samples (about 300 g) were separated for 20 minutes in a sieving machine with shower. The median grain size was calculated and expressed as a percentage of total weight. To determine the organic content, about 10 g of dried samples were ashed at 55°C for 2 hours, then weighed.

The epifauna was collected semi-quantitatively using a 25 cm squaremouthed hand-dredger. At the same station, the infauna was collected by

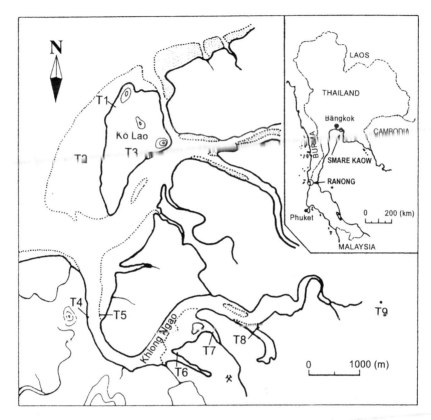

Figure 5.1 Sketch of Thailand, showing location of study sites (and transects T1–T9) at Smare Kaow and Ranong. Source: after Shokita et al. (1983)

digging down into the soil to a depth of 15–20 cm, then sieving the soil with a sieve, mesh size 1 mm × 1 mm.

The zonal distribution and abundance of macrobenthos in Thai mangals, based on Shokita et al. (1983), are shown in figures 5.2–5.4.

Results at Smare Kaow mangal

A total of 20 epifaunal and infaunal species was found at the sampling stations. The tiny tanaidacean *Apseudes* sp. was most abundant at stations 2 and 3, accounting for 73.2 per cent of all species. The next most abundant species was the lamellibranch mollusc *Glauconome chinensis* (11.9 per cent). At the seaward edge of station 4, species living around the mid-zone were absent and were replaced by the bivalves *Moerella* sp. and *Cuspidaria* sp., the tiny gastropod *Stenothyra* sp., and the polychaete

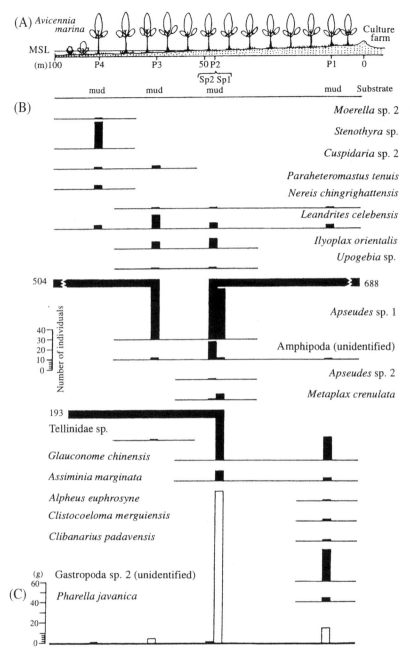

Figure 5.2 Zonation and abundance of macrofauna at Smare Kaow: (A) schematic profile; (B) distribution and abundance of macrofauna per 625 cm^2; (C) standing stock (wet weight/625 cm^2); MSL, mean sea level. Source: after Shokita et al. (1983)

Nereis chingrighattensis. Substation 2 had the highest density, with *G. chinensis* being dominant (fig. 5.2).

Results at the Ranong mangal

The transects are shown in figure 5.1: the results at transects 1, 3, and 6 are as follows.

Transect 1 (200 m long), located at the seaward edge facing the Andaman Sea, north of Ko Lao, had six sampling points (fig. 5.3). The mangrove species consisted primarily of *Avicennia marina* and *Rhizophora* spp. The soil was muddy, the median diameter of particles was 0.05–0.06 mm, and the silt-clay content was 49–70.4 per cent. The soil was high in organic content, except at subpoint 2 in the open area of the mangrove forest.

Altogether, 21 species were found: these comprised polychaetes (6), molluscs (6), crustaceans (7), a sipunculid worm, and an unidentified ophiuroid. Figure 5.3 shows the polychaetes, ophiuroid, some species of crustaceans – such as *Diogenes avarus*, *Alpheus euphrosyne*, and *Macrophthalmus crinitus* – and the gastropod mollusc *Paratectonatica tigrina* that were found in the mud-flat. The barnacle *Balanus amphitrite* and the gastropod mollusc *Clypeomorus patulum* were abundant species that occurred near the border of the mangrove forest and mud-flat: the former was found on stones and on the prop roots and trunks of mangrove trees. Other species found in the biotope were the fiddler crab *Uca forcipata*; the hermit crab *Clibanarius padavensis*; the grapsid crab *Metopograpsus oceanicus*; the gastropod molluscs *Cymia gradata*, *Cerithideopsilla djadjariensis*, and *Rhizophomurex capucinus*; the bivalve *Eamesiella corrugata*; and the sipunculid worm *Phascolosoma arcuatum*. *Cerithideopsilla djadjuriensis* was the most abundant of all species collected. The raw weight of the macrofauna was highest at subpoint 1 because of the shell weight of *R. capucinus*. Generally, the animal biomass was higher in the mangrove biotope than in the mud-flat.

Transect 3 (130 m) at the channel of the east coast of Ko Lao had five sampling points, two of which were on the mud-flat and three in the mangrove forest. The flora of this area consisted of *Rhizophora* spp., *Bruguiera* spp., and a sparse population of *Acanthus ilicifolius* (fig. 5.4).

The soil was generally muddy; point 5 was semifluid with silt-clay (87 per cent), but point 1 in the mangrove forest was relatively sandy. The soil was high in organic matter (11.0–22.8 per cent). Point 2 was rich in organisms and had a dense population of *Bruguiera* sp.

In all, 28 macrobenthos species were found in the mud-flat (17) and in the mangrove forest (17). They consisted of polychaetes (5), molluscs (13), crustaceans (8), a sipunculid worm, and an unidentifed ophiuroid. Molluscs were predominant: point 5 in the mud-flat had the greatest

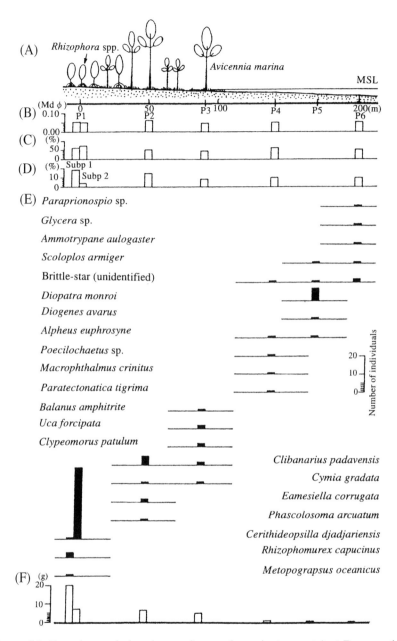

Figure 5.3 Zonation and abundance of macrofauna in transect 1 at Ranong: (A) schematic profile; (B) median diameter; (C) silt-clay content; (D) organic content; (E) distribution and abundance of macrofauna per 625 cm^2; (F) standing stock (wet weight/625 cm^2); MSL, mean sea level. Source: after Shokita et al. (1983)

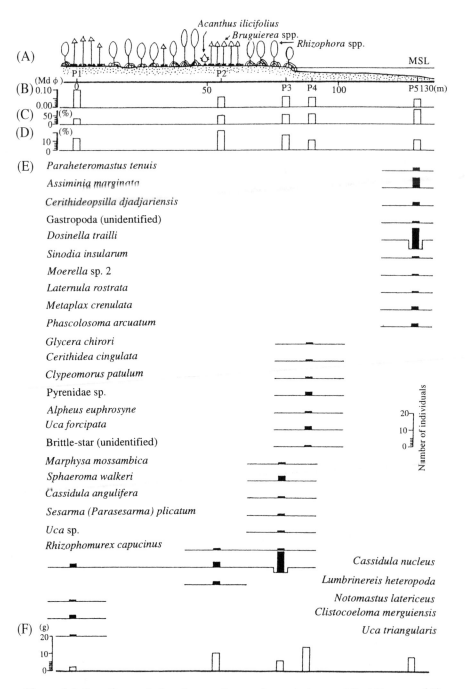

Figure 5.4 Zonation and abundance of macrofauna in transect 3 at Ranong: (A) schematic profile; (B) median diameter; (C) silt-clay content; (D) organic content; (E) distribution and abundance of macrofauna per 625 cm^2; (F) standing stock (wet weight/625 cm^2); MSL, mean sea level. Source: after Shokita et al. (1983)

number of species, the small infauna lamellibranch *Dosinella trailli* being particularly abundant. The gastropod *Cassidula nucleus* was dominant in the mangrove forest. When the ebb tide started, this species immediately migrated from the trunk to the mangrove floor to feed on organisms in the mud; at the start of the flood tide, this snail again ascended the mangrove trunks. The animal total biomass was comparatively high at points 2 and 4.

Transect 6 (120 m) was drawn across the mud-flat and channel from the *Sonneratia* forest of the right bank to the mixed mangrove forest of the left bank. Six sampling points were established. The forest of the left bank consisted of *S. alba*, *Rhizophora* spp., and *Bruguiera* spp. The soil of this area was muddy with silt-clay and the organic content was relatively high (fig. 5.5).

In all, 29 species were found: these comprised an unidentified nemertinean, polychaetes (4), molluscs (8), crustaceans (15), and a sipunculid worm. Point 6 had a high species diversity and the highest animal biomass. At point 3, large numbers of the gastropod *Assiminia marginata* moved around the pneumatophores of *S. alba*. The animal biomass was greatest on the right bank of the forest.

Macrofaunal habitats and dominant species

The mangal areas are transitional between land and sea. Mangal-dwelling animals are derived from either realm, and there are different patterns of animal zonation. Berry (1964) recognized five zonation patterns in the Malayan mangals, namely (1) *Littorina*, (2) *Nerita*, (3) bivalves, (4) *Uca*, and (5) burrower zones.

The *Littorina* zone corresponds essentially to that described by Stephenson and Stephenson (1949) from other types of shores and is characterized by the presence of *Littorina* species. The *Nerita* zone is situated in the area of lower mangrove trees and is characterized by the presence of other gastropods, *Nerita* species in particular.

The bivalve zone is situated at the seaward fringe and is characterized by the presence of bivalves that do not occur in the forest. The *Uca* zone is situated at the muddy–sandy surface that is rich in crabs, especially *Uca* species. These crabs, unlike most other crabs, do not leave the mud surface and can be seen in large numbers between the mid-tide level and high-water mark at neap tide. The burrower zone is located on the coast or on channel edges.

On the basis of Berry's study (Berry 1964) and bearing in mind the surrounding environment, the mangal ecosystem may be divided into the following four biotopes: (1) landward edge; (2) mangal proper; (3) outer-mangal muddy–sandy tidal flat; (4) channel or embayment biotopes. The

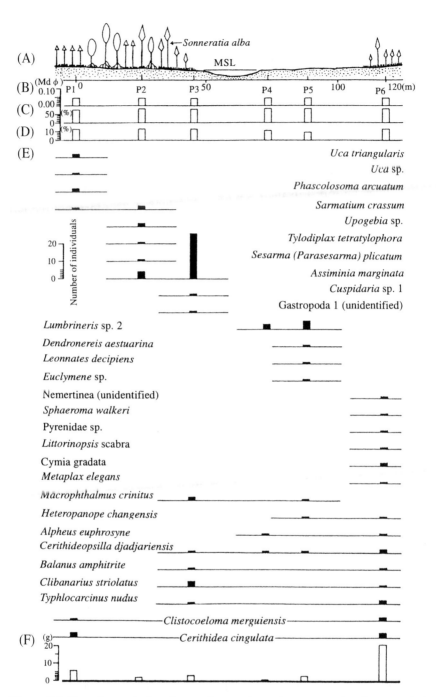

Figure 5.5 Zonation and abundance of macrofauna in transect 6 at Ranong: (A) schematic profile; (B) median diameter; (C) silt-clay content; (D) organic content; (E) distribution and abundance of macrofauna per 625 cm^2; (F) standing stock (wet weight/625 cm^2); MSL, mean sea level. Source: after Shokita et al. (1983)

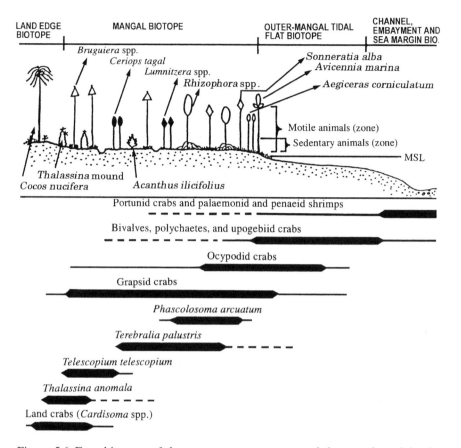

Figure 5.6 Four biotopes of the mangrove ecosystem, and the zonation of dominant groups and species of the macrofauna in Thailand. Source: after Shokita (1989)

macrobenthic zonation of these biotopes in the Thai mangal is shown in figure 5.6 (Shokita 1989).

Table 5.2 indicates that the mangal proper and the outer-mangal flat biotopes may be further subdivided into seven and two habitats, respectively. This table shows the representative animals living in each habitat of the Okinawan and Thai mangals.

Tree-dwelling fauna (photos 5.3, 5.4)

Tree-dwelling animals abound in the mangroves at the seaward edge and along the channels. They may be sedentary or motile species. Members

Table 5.2 Macrofaunal habitats and representative animals on Okinawan and Thai mangals

Biotopes			Representative animals
I	Landward edge		*Cardisoma carnifex*, *Caenobita cavipes*, *Holometopus dehaani*
II	Mangal proper	surface	*Terebralia palustris*, *Telescopium telescopium*, *Cerithidea obtusa*
		inner	*Thalassina anomala*, *Phascolosoma arcuatum*, *Alpheus* spp.
		rivulet	*Leandrites celebensis*, *Macrobrachium equidens*, *Metapenaeus* spp., *Yongeichtys criniger*
		tide-pool	Gobiid fish, *Caridina* spp., *Palaemon debilis*, *M. equidense*
		sea or river boundary	*Uca* spp., *Clipeomorus patulum*, *Rhizophomurex capucinus*
		tree: sedentary	*Saccostrea cucullata*, *Merina ephippium*, *Hormomya mutabilis*, *Balanus amphitrite*
		motile	*Tittorinopsis scabra*, *Nerita* spp., *Cassidula* spp., *Metopograpsus oceanicus*
		dead tree	*Beteropanope changensis*, Blennies
III	Outer-mangal flat	surface	*Uca* spp., *Macrophthalmus* spp., *Metaplax crenulata*, *Diogenes avarus*, *Boleophthalmus boddaerti*
		inner	*Nereis chingrighattensis*, *Diopatra monroi*, *Glauconome chinensis*, *Paphia luzonica*
IV	Channel or embayment		*Acetes erythraeus*, *Metapenaeus* spp., *Penaeus* spp., *Thalamita crenata*, *Scylla serrata*, *Mugil cephalus*, *Siganus guttatus*, *Acanthopagurus silvicolus*, *Carcharhinus leucus*

Source: Modified from Shokita et al. (1983).

of the former species include the barnacles *Balanus amphitrite*, *B. albicostatus*, and *Europhia withersi*; the bivalves *Melina ephippium*, *Hormomya mutabilis*, *Modiolus metcalfei*, *Saccostrea cucullata*, *Crassostrea vitrefacta*, and *Trapezium (Neotrapezium) liratum*. The latter (motile) species are the gastropods *Littorina scabra*, *L. melanostoma*, *L. carinifera*, *Nerita articulata*, *N. (Ritena) undulata*, *N. (Theliosyla) planospira*, *Dosnia violacea*, *Cymia gradata*, *Rhizophomurex capucinus*, *Cassidula mustelia*,

C. nucleus, C. angulifera, and *Cerithidea rhizophorarum marchii*; the grapsid crabs *Metopograpsus latifrons, M. oceanicus, Chiromantes bidens, Neoepisesarma mederi, N. lafondi*; and the mud skipper *Periophthalmus vulgaris* (MacNae 1968; Plaziat 1984; Shokita 1989).

Table 5.3 compares the composition and abundance of tree-dwelling fauna in subtropical Okinawan and tropical Thai mangals. The mangals of Thailand have a richer fauna than those of Okinawa, possibly because the latter consist of smaller communities. Lower winter temperatures also may prevent the survival of tropical species (Shokita 1989).

Stock enhancement of economically important species

Many economically important species occur in the mangrove areas of the Indo-Pacific, such as the penaeid shrimp *Penaeus monodon*, the mud crab *Scylla* spp., the milkfish *Chanos chanos*, the rabbitfish *Siganus guttatus*, the black bream *Acanthopagrus silvicolus*, the snapper *Lutjanus argentimaculatus*, and the barramundi *Lates calcarifer*. The numbers of these species are decreasing in the Indo-Pacific, owing to overfishing. The exception is *Scylla* spp., which shows trends of stock increase in Japan. *Scylla* is an economically important Indo-Pacific genus but the taxonomy of the species is confusing. Keenan, Davie, and Mann (1998), who collected specimens from many localities throughout the Indo-Pacific region and studied them from various biological viewpoints, recognized the following four species: *S. serrata, S. olivacea, S. paramamosin*, and *S. tranquebarica*. All species except *S. tranquebarica* are found in Japan.

Wild populations of these crabs in the Indo-Pacific tend to decrease yearly as a result of overfishing. To enhance the population of mangrove crabs in Japan, a large artificial seed production plant of *S. serrata* was developed several years ago at the Yaeyama Station of the Japan Seafarming Association. The larvae are released into the natural environment for stock enhancement, representing one aspect of marine ranching (K. Hamazaki, Jap. Sea-farm. Assoc, pers. com.).

Feeding habits

Feeding habits of molluscs

Six types of feeding habits of mangal molluscs are recognized: these are (1) suspension feeders, (2) filter feeders, (3) herbivores and deposit feeders, (4) carnivores, (5) deposit feeders–mud raspers, and (6) algae and detritus feeders (Plaziat 1984).

Most bivalves are aquatic suspension feeders whose vertical distribution does not extend above the mean high-water neap-tide level. On mud-flats, filter feeders appear only very rarely in the upper intertidal

Table 5.3 Comparison of species composition and abundance of tree fauna between Okinawan and Thai mangals

Species	Okinawan (Iriomote I.)	Thailand (Ranong)
I. SEDENTARY		
A. Cirripeda		
1. *Balanus amphitrite*	0	+++
2. *B. albicristatus*	++	0
3. *Euraphla withesi*	0	++
B. Bivalvia		
4. *Melina ephippium*	−	+++
5. *Hormomya mutabilis*	+	+++
6. *Modiolus metcalfei*	0	+
7. *Saccostrea cucullata*	0	+++
8. *Crassostrea vitrefacta*	+	0
No. of species	4	6
II. MOTILE		
C. Gastropoda		
1. *Littorinopsis scabra*	++	+++
2. *L. melanostoma*	0	+
3. *L. carinifera*	0	++
4. *Nerita articulata*	0	+++
5. *Nerita (Retina) undulata*	+	0
6. *N. (Theliostyla) planospira*	+	+
7. *Dostia violacea*	+	++
8. *Cymia gradata*	0	++
9. *Rhizophomurex capucinus*	0	+++
10. *Cassidula mustelina*	0	+
11. *C. nucleus*	−	++
12. *C. angrifera*	0	+
13. *Cerithidea rhizophorarum*	++	0
D. Decapoda		
14. *Metopograpsus latifrons*	+	0
15. *M. oceanicus*	0	+
16. *Chiromantes bidens*	+++	0
17. *Neoepisesarma lafondi*	++	0
18. *N. mederi*	0	++
E. Fish species		
19. *Periophthalmus vulgaris*	+++	+++
No. of species	10	14
Total	14 (41.2%)	20 (58.8%)

Source: Shokita et al. (1983).
Abbreviations: 0, not found; −, very few; +, common; ++, fairly abundant; +++, very abundant.

zone, one exception being the *Polymesoda* species of New Caledonia (Plaziat 1984). Tree-dwelling suspension feeders tend to occupy the seaward edge of mangroves; Berry (1964) has suggested that this may be due to the need for planktonic food from the open sea.

In the Thai mangals, the carnivorous snails *Cymia gradata* and *Rhizophomurex capucinus* feed on the oyster *Saccostrea cucullata* and on barnacles. *R. capucinus* also feeds on the young of the potamodid snail *Terebralia palustris* at the Ao Khung Kraben mangal, which is located at the Chanthaburi Province of eastern Thailand (Shokita et al. 1983).

Plaziat (1977), Nishihira (1983), and Nishihira, Tsuchiya, and Kubo (1988) observed that the giant snail *T. palustris* directly grazes on mangrove litter such as leaves, stipules, calyces, fruits, and propagules, as well as on detritus. Among these, leaf litter was the most frequent food item. This snail grazes naturally not only on fallen leaves but also on fresh green leaves that have fallen to the ground, whereas young snails of this species are deposit feeders. This species may play an important role in the degradation of mangrove litter *in situ*.

Feeding habits of crabs

Mangal-dwelling crabs were divided into seven groups, according to their feeding habits, by Jones (1984) – namely herbivores, carnivores, omnivores, deposit feeders, omnivore/deposit feeders, specialized filterers, and omnivore/filterers. Table 5.4 shows the food preference and feeding habits of Okinawan mangal crabs (Shokita et al. 1998). The ocypodid crabs are generally deposit feeders or detritivores, feeding on meiofauna, microscopic algae, and organic matter on the surface of the substratum and other solid matter. Many grapsid crabs are herbivores and omnivore/deposit feeders, scraping mangrove litter, water plants, organic deposits, and carrion. The portunid and a few grapsid crabs are predominantly carnivores, feeding on molluscs, shrimps, crabs, and fish.

Mangrove litter is ingested directly by many herbivorous/omnivorous crabs, such as the grapsid crabs *Helice leachi*, *Perisesarma guttata*, *P. onychophorum*, *P. dussumieri*, *P. semperi*, *P. eumolpe*, *P. bidens*, *Neosarmatium smithi*, *N. meinerti*, *Neoepisesarma (N.) lafondi*, *N. (N.) mederi*, *N. (N.) versicolor*, *Chiromanthes dehaani*, *C. maipoensis*, *Parasesarma lanchesteri*, *Metopograpsus latifrons*, *M. messor*, and *Aratus pisoni*; the xanthid crab *Rhithropanopeus harrisii*; and the land crabs *Cardisoma carnifex*, *Ucides cordatus*, and *U. occidentalis* (Camilleri 1992; Giddens et al. 1986; Jones 1984; Lee 1989; MacNae 1968; Mia et al. 1999; Nakasone and Agena 1984; Nakasone, Limskul, and Tirsrisook 1985; Odum and Heald 1975; Robertson 1986, 1988; Robertson and Daniel 1989; Twilley et al. 1997; Warner 1977). Almost all of these crabs feed

Table 5.4 Crustacean species found at the mangrove swamp of the Okukubi River in 1979 and 1997, showing feeding habits and habitat

Species	Feeding habits	Habitat	Year	
			1979	1997
Family Penaeidae				
Metapenaeus moyebi (Kishinouye, 1896)	Omnivore	Channel	✓	✓
Family Palaemonidae				
Palaemon (P.) debilis Dana, 1852	Omnivore	Channel	✓	✓
P. (P.) concinnus Dana, 1852	Omnivore	Channel	✓	
P. (P.) macrodactylus Rathbun, 1902	Omnivore	Channel		✓
Macrobrachium formosense Bate, 1868	Omnivore	Channel		
Family Hippolytidae				
Merguia oligodon De Man	Omnivore			✓
Family Alpheidae				
Alpheus sp.	Omnivore	Stony area		✓
Family Laomediidae				
Laomedia astacina De Haan, 1849	Deposit feeder	Forest	✓	
Family Thalassinidae				
Thalassina anomala (Herbst, 1804)	Deposit feeder	Forest	✓	✓
Family Diogenidae				
Clibanarius longitarsus (De Haan, 1849)	Omnivore	Channel	✓	✓
Family Coenobitidae				
Coenobita rugosus H. Milne Edwards, 1873	Omnivore	Landward edge	✓	✓
C. cavipes Stimpson, 1858	Omnivore	Landward edge	✓	
Family Portunidae				
Scylla oceanica Estampador, 1949	Carnivore	Channel and forest	✓	
Portunus (P.) pelagicus (Linnaeus, 1758)	Carnivore	Channel	✓	
Thalamita crenata (Latreille, 1829)	Carnivore	Channel	✓	✓

Species	Feeding	Habitat	✓
Family Xanthidae			
Leptodius exaratus (H. Milne Edwards, 1834)		Channel and forest	✓
L. leptodon Forest & Guinot, 1961		Stony area	✓
Baptozium vinosus (H. Milne Edwards, 1834)	Carnivore	Forest and stony area	✓
Chlorodiella nigra (Forskål, 1775)		Stony area	✓
Epixanthus dentatus (White, 1847)	Carnivore	Stony area	✓
Pilumnopeus marginatus (Stimpson, 1858)		Stony area	✓
P. makianus (Rathbun, 1929)		Stony area	
Family Gecarcinidae			
Cardisoma carnifex (Herbst, 1796)	Omnivore	Landward edge	✓
Family Grapsidae			
Metopograpsus thukuhar (Owen, 1839)	Omnivore	Stony area	✓
M. latifrons (White, 1848)	Omnivore	Forest and stony area	✓
Psychognathus glaber Stimpson, 1958	Omnivore	Outer-mangal flat	✓
Varuna litterata (Fabricius, 1798)	Omnivore	Channel	✓
Chiromantes dehaani (H. Milne Edwards, 1853)	Omnivore	Landward edge	✓
Parasesarma plicatum (Latreille, 1803)	Omnivore	Landward edge	✓
P. pictum (de Haan, 1835)	Omnivore	Landward edge	✓
P. acis Davie, 1993	Omnivore	Landward edge	✓
Perisesarma bidens (De Haan, 1835)	Omnivore	Forest	✓
Neoepisesarma (N.) lafondi (Jaquinot & Lucas, 1853)	Omnivore	Forest	✓
Chasmagnatus convexus De Haan, 1833	Omnivore	Landward edge	✓
Helice leachi Hess, 1865	Herbivore	Forest	✓
H. formosensis Rathbun, 1931	Carnivore	Forest and outer-mangal flat	✓
Pyxidognathus sp.			✓
Family Mictyridae			
Mictyris brevidactylus Stimpson, 1858	Deposit feeder	Outer-mangal flat	✓
Family Ocypodidae			
Ocypode ceratophthalma (Pallas, 1772)	Omnivore	Strand	✓
O. cordimana Desmarest, 1825	Omnivore	Strand	✓

Table 5.4 (cont.)

Species	Feeding habits	Habitat	Year 1979	Year 1997
Uca (Deltuca) dussumieri dussumieri (H. Milne Edwards, 1852)	Deposit feeder	Forest and outer-mangal flat	✓	
U. (Celuca) lactea perplexa (H. Milne Edwards, 1837)	Deposit feeder	Forest and outer-mangal flat	✓	✓
U. (Amphiuca) chlorophthalma crassipes (Adams & White, 1848)	Deposit feeder	Forest and outer-mangal flat	✓	✓
U. (Thalassuca) vocans vocans (Linnaeus, 1758)	Deposit feeder	Forest and outer-mangal flat	✓	✓
Scopimera globosa (De Haan, 1835)	Deposit feeder	Outer-mangal flat	✓	✓
Macrophthalmus (M.) verreauxi H. Milne Edwards, 1848	Deposit feeder	Channel	✓	✓
M. (M.) convexus Stimpson, 1898	Deposit feeder	Channel and outer-mangal flat		✓
M. (M.) definitus White, 1848	Deposit feeder	Channel and outer-mangal flat	✓	✓
M. banzai Wada & Sakai, 1989	Deposit feeder	Channel and outer-mangal flat	✓	
Total			39	29

Source: Modified from Shokita et al. (1998).

not only on mangrove leaves directly but also on the terrestrial vegetation and seedlings, calyx, fruits, and stipules of mangroves.

Feeding habits of Helice *species*

Helice leachi is a typical herbivorous crab, consuming mangrove litter directly on the forest floor. The results of studies of the feeding habits of this crab from Okinawa Island are given below. It is abundant in the mangrove forests and the grassy, upper tidal areas near estuaries and inland seas. This species is distributed throughout Japan, Korea, Taiwan, Indonesia, Australia, Madagascar, the east coast of Africa, Caroline Islands, New Caledonia, the Moluccas, and Lombok (Dai and Yang 1991; Miyake 1983; Sakai 1976).

Mia et al. (1999) studied the preference of *H. leachi* for different types of mangrove leaves. Their experiments investigated whether *H. leachi* can survive feeding on leaves alone, which type of leaf they prefer, and their food choice.

Experiment 1

Bruguiera gymnorrhiza leaves were divided into three categories – green, yellow, and brown. Different types of leaves were given to individual crabs in separate containers. To investigate the leaf-processing rate, the leaves were weighed daily to the nearest 0.001 g using an electric balance. New leaves were added to each container when less than 0.03 g of the previous leaf remained. All leaves were photocopied every day to determine the mode of leaf processing.

During the experimental period, 100 per cent of *H. leachi* survived, showing that these crabs can live for at least six weeks by eating only green, yellow, or brown *B. gymnorrhiza* leaves. The crabs first cut the leaves using their chelae, then shred them using their mandibles. The average weekly consumption rates of brown, yellow, and green leaves were 0.63, 0.22, and 0.33 g per gram of *H. leachi* (table 5.5). The rates were calculated as the weight differences between perfect and consumed leaves. The consumption rate of brown leaves was higher than that of green or yellow leaves, indicating a preference for brown leaves.

Experiment 2

Green, yellow, and brown leaves of *B. gymnorrhiza* were provided simultaneously to assess the choice preference. The experiment was repeated six times. The growth of the crabs and leaf processing were monitored as described in experiment 1.

When individuals of *H. leachi* were given a choice, they selected brown, rather than green or yellow, leaves of *Bruguiera gymnorrhiza*. During the course of the experiment, the consumption rate of brown

Table 5.5 Average weekly rates of leaf consumption by *Helice leachi* per gram body weight when provided with green, yellow, and brown leaves of *Bruguiera gymnorrhiza* separately

Week no.	Consumption rates (g)		
	Green	Yellow	Brown
1	0.29	0.02	0.19
2	0.30	0.48	0.71
3	0.28	0.47	0.87
4	0.21	0.07	0.42
5	0.58	0.00	0.94
Average	0.33	0.22	0.63

Source: After Mia et al. (1999).

leaves was higher than that of other leaves. The average weekly consumption rate of brown, yellow, and green leaves was 0.31, 0.18, and 0.27 g, respectively, per gram of *H. leachi* (table 5.6). The order of leaf preference was brown, green, and (least of all) yellow, a result consistent with the results of the previous experiment.

Since none of the crabs died during either of these experiments and their growth was not hampered, *H. leachi* is regarded as primarily herbivorous.

Chemical analysis of mangrove leaves

Because *H. leachi* can survive by eating various leaves of *B. gymnorrhiza*, the nutritional composition of each type of these leaves was analysed.

Table 5.6 Average weekly rates of leaf consumption by *Helice leachi* per gram body weight when provided with green, yellow, and brown leaves of *Bruguiera gymnorrhiza* together

Week no.	Consumption rates (g)		
	Green	Yellow	Brown
1	0.04	0.04	0.44
2	0.10	0.08	0.35
3	0.26	0.22	0.42
4	0.42	0.26	0.38
5	0.27	0.16	0.13
6	0.54	0.32	0.12
Average	0.27	0.18	0.31

Source: After Mia et al. (1999).

Table 5.7 Nutritional composition of green, yellow, and brown leaves of *Bruguiera gymnorrhiza* per 100 g wet weight

Composition	Green	Yellow	Brown
Energy (kcal)	91	98	118
Water (g)	71.4	70.1	65.1
Protein (g)	2.1	0.6	1.0
Fat (g)	0.7	1.4	2.0
Carbohydrate (g)	22.3	23.8	27.8
Ash (g)	3.5	4.1	4.1

Source: After Mia et al. (1999).

The total energy, water, protein, fat, carbohydrates, and ash contents of 100 g of wet and dry green, yellow, and brown leaves of *B. gymnorrhiza* are given in tables 5.7 and 5.8. Dry leaves were more nutritious than wet leaves. The contents of energy, fat, and carbohydrates were higher in both wet and dry brown leaves than in yellow or green leaves. Among wet and dry leaves of all colours, the green leaves contained more protein.

Preference of *Helice leachi* for brown leaves

Experiments showed that these crabs can live on green, yellow, and brown leaves of *B. gymnorrhiza* and start feeding from any parts of the leaf. They preferred brown, to green and yellow aged leaves, and the consumption rates of green and yellow leaves were lower than those of brown leaves. These results are consistent with those of Giddens et al. (1986), who found that the grapsid crab *Neosarmatium smithi* consumed decayed leaves of *Ceriops tagal* in preference to fresh or senescent leaves. Micheli (1993) noted that, in laboratory experiments on selective feeding, *Sesarma messa* and *N. smithi* did not show any preference for

Table 5.8 Nutritional composition of green, yellow, and brown leaves of *Bruguiera gymnorrhiza* per 100 g dry weight

Composition	Green	Yellow	Brown
Energy (kcal)	317	328	339
Water (g)	–	–	–
Protein (g)	7.3	2.0	2.9
Fat (g)	2.5	4.7	5.7
Carbohydrate (g)	78.0	79.6	79.7
Ash (g)	12.2	13.7	11.7

Source: After Mia et al. (1999).

recently fallen leaves of four species of mangroves (*Avicennia marina*, *R. stylosa*, *B. exaristata*, and *C. tagal*), and that they consumed significantly more decayed than senescent leaves. Three hypotheses have been offered to explain the preference for aged, over freshly fallen, leaves:

1. The first is that unpalatable materials are removed from the leaf during the process of ageing – namely, tannins, which may gradually leach out of leaves into sea water. Tannins are polyphenolic compounds that bind to digestive enzymes, decreasing absorption efficiency; they deter herbivory by various grazers. Mangrove leaves that have just fallen from trees are poor in nitrogen and rich in tannins. During leaching the tannin content decreases; therefore, brown (aged) leaves of *B. gymnorrhiza* would probably contain the lowest amount of tannin of all leaf types offered. This may, in part, account for the preference for brown leaves.
2. The second hypothesis is that chemicals produced in ageing leaves may act as an attractant to crabs. The fungal colonization of plant material affects its flavour. This may provide shredders with feeding cues, if such species have appropriate chemoreceptors.
3. The third hypothesis is that the nutritional quality of leaves improves during ageing; this, in turn, could be due to gradual colonization of leaves by fungi. This hypothesis is supported by the observation that crustaceans prefer leaves that have been inoculated with fungi, and use the fungi either as a source of energy or as an aid in the digestion of plant material. As a result of fungal colonization, the absolute nitrogen content of dead leaves may increase, thus implying that the crustacean preference for microbially infected leaves is based on nutritional quality. The nitrogen content of *R. mangle* leaves increased during ageing.

Helice leachi plays an important role in the energy-flow pathway in the mangrove forest by providing food for detritus feeders such as crustaceans, molluscs, chironomid larvae, and nematodes. Odum (1971) also found that crab faecal pellets are important dietary components for three common fishes. The nitrogen concentration in decaying brown leaves is increased in both freshwater and estuarine food webs and reflects the formation of microbial protein. The tissues of green leaves contain more water because of the soft layer. The present study showed that brown leaves contained more fat than yellow or green leaves.

In the diverse mangrove systems of Okinawa, South-East Asia, and Australia, sesarmid crabs graze on a large scale on mangrove litter, consistently reducing the tidal export rate and accelerating the litter breakdown. In north Queensland, Australia, 28–79 per cent of the annual litter fall is removed by sesarmid crabs in the low and mid–low intertidal *Rhizophora* forests (Robertson 1986, 1988; Robertson and Daniel 1989). In

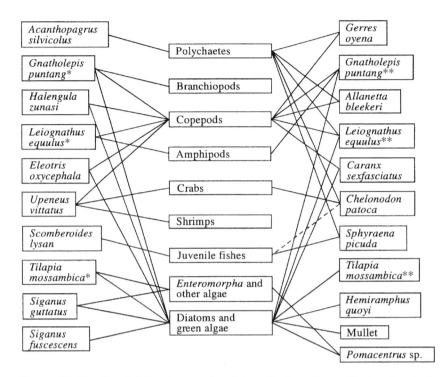

Figure 5.7 Feeding habits of estuarine fishes. Broken line indicates direct observation; single and double asterisks indicate fishes with occasional seasonal changes in food preference. Source: after Nakasone and Agena (1984)

high intertidal forests, the removal rates of litter by crabs may reach 79 per cent.

Feeding habits of fish species

Nakasone and Agena (1984) and Shokita et al. (1988) analysed the stomach content of many fishes collected from estuarine waters in Okinawa and the Yaeyama islands. The foods of estuarine fishes are varied: they include polychaetes, various taxa of crustaceans, fishes, and various algae (fig. 5.7). Some fishes change feeding habits with the season. Fishes with longer intestines tend to feed on plant material; those with short intestines tend to feed on animal material.

The feeding habits of fish species associated with the mangals are divided into (1) herbivores (*Siganus guttatus* and *Oreochromis mossambicus*), (2) omnivores (*Mugil cephalus* and *Nematalosa come*), (3) carnivores (*Sphyraena barracuda* and *Carcharinus leucus*), and (4) zooplanktivores (*Hypoatherina woodwardi*). Carnivores are subdivided into fish eaters, such as *S. barracuda* and *C. leucus*, and benthos eaters, such as *Gerres* species and *Hapalogenys nitens*.

Food chain and role of aquatic animals in degradation of mangrove litter

Animals of mangrove swamps are divided chiefly into three types according to their habits. Those in the first group live in the mangroves throughout their lives; those in the second group live in the mangroves only as juveniles and young adults; animals of the third group forage in the swamp both as young and as adults.

Figure 5.8 depicts an example of a food chain in a mangrove swamp on Iriomote Island (Shokita 1991). Energy flows through two major pathways in the food web. In the first, mangrove litter is directly consumed by crabs, gastropods, and others; then, large carnivorous animals, such as mangrove crabs, and fish, such as *Acanthopagrus* species, eat them in turn. In the second pathway, litter is decomposed into detritus by bacteria and fungi and is then eaten by the meiobenthos, detritus feeders, and zooplankton, which are subsequently consumed by larger carnivorous fishes.

Leaf material ingested by grapsid crabs and by the potamodid snail *T. palustris* is incompletely digested and most of it is returned to the environment as large faecal pellets that contain more finely divided leaf material. Such macrobenthos plays an important role, as it breaks down mangrove litter into detrital particles and hastens the decomposition process. A large number of grazing animals – such as amphipods, isopods, shrimps, crabs, and snails – are associated with the mangrove forest. Crabs make food available to other species by shredding leaves into smaller particles (Nakasone and Agena 1984; Nakasone, Limsakul, and Tirsrisook 1985).

Rivers also bring large quantities of organic particulate matter, litter, algae, juvenile fish, and freshwater insects to mangrove areas and this serves as food for detritus feeders. Moreover, fragments of algae, sea grasses, and dead animals are washed into mangrove areas from the sea (Shokita 1991). The four basic trophic groups of the food chain in mangroves are detrivores, herbivores, omnivores, and carnivores.

Mangroves and sandy–muddy flats are utilized during flood tide by many periodic foragers (shrimps, crabs, fishes, and birds). The abundance of macro- and meiofauna, phytoplankton, zooplankton, and detritus in the mangroves provides a large store of food for visitors and residents.

Resident and migratory fish feed on detritus, plankton, benthos, annelids, gastropods, bivalves, invertebrate larvae, insects, mysids, amphipods, isopods, copepods, and on the larvae and juveniles of shrimps,

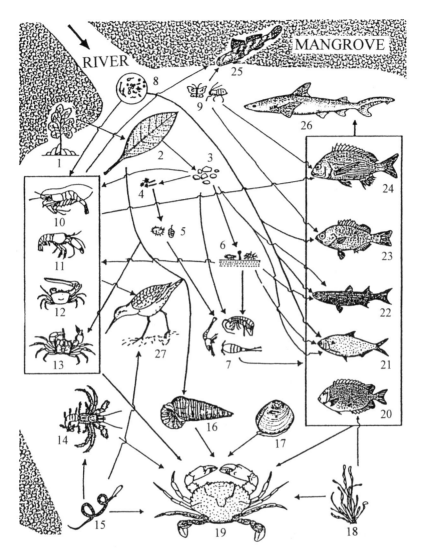

Figure 5.8 Food chains in mangroves of Iriomote Island, Okinawa: (1) mangrove tree (*Rhizophora stylosa*); (2) mangrove litter; (3) detritus; (4) bacteria; (5) protozoans; (6) meiobenthos; (7) zooplankton; (8) organic matter from river; (9) insects; (10) shrimp (*Palaemon debilis*); (11) snapping shrimp (*Alpheus* spp.); (12) fiddler crab (*Uca* spp.); (13) grapsid crab (*Helice leachi*); (14) hermit crab (*Clibanarius longitarsus*); (15) polychaetes; (16) potamodid snail (*Terebralia palustris*); (17) bivalve (*Geloina erosa*); (18) seaweed; (19) mud crab (*Scylla* spp.); (20) rabbitfish (*Siganus guttatus*); (21) gizzard shad (*Nematalosa come*); (22) striped mullet (*Mugil cephalus*); (23) flagtail (*Kuhla* spp.); (24) black bream (*Acanthopagrus silvicolus*); (25) mud skipper (*Periophthalmus vulgaris*); (26) shark (*Carcharhinus leucus*); (27) phalarope. Source: after Shokita (1991)

crabs, and fishes that are abundant in the mangroves. Food resources in the mangrove support many organisms such as shrimp, crabs, mullets, groupers, snappers, croakers, shads, sea perches, catfishes, and anchovies.

Conclusions

Mangrove areas are important feeding grounds for many coastal-water fishes and serve as good nurseries for juvenile fishes and crustaceans. They attract and shelter fishes under the shade of the mangrove forest. To maintain and enhance fishery resources in mangrove areas, the natural environment of mangrove communities must be protected from human disturbance. Mangroves must be planted in tidal flats, which will form fishing grounds and numerous new habitats for many marine organisms, including migratory fishes.

Note

1. Throughout this volume, photographs have been placed at the end of the chapter to which they refer.

REFERENCES

Berry, A.J. 1964. "Faunal zonation in mangrove swamps." *Bulletin of the National Museum of Singapore* 32: 90–98.

Camilleri, J. 1992. "Leaf choice by crustaceans in a mangrove forest in Queensland." *Marine Biology* 102: 453–459.

Dai, A.Y. and S.L. Yang. 1991. *Crabs of the China Seas*. Beijing: China Ocean Press, 682 pp.

Day, Jr., J.W., C.A.S. Hall, W.M. Kemp, and A. Yànez-Arancibia. 1987. *Estuarine Ecology*. New York: John Wiley and Sons, 558 pp.

Frith, J.W., R. Tantanasiriwong, and O. Bhatia. 1976. "Zonation of macrofauna on a mangrove shore, Phuket Island, southern Thailand." *Phuket Marine Biology Center Research Bulletin* 17: 1–14.

Giddens, R.L., J.S. Lucas, M.J. Neilson, and G.N. Richards. 1986. "Feeding ecology of the mangrove crab, *Neosarmatium smithi* (Crustacea: Decapoda: Sesarmidae)." *Marine Ecology Progress Series* 33: 147–155.

Gunderman, N. and D.M. Popper. 1984. "Notes on the Indo Pacific mangal fishes and on mangrove related fisheries." In: F.D. Por and I. Dor (eds). *Hydrobiology of the Mangal*. The Hague: Dr W. Junk Publishers, 201–209.

Jones, D.M. 1984. "Crabs of the mangal ecosystem." In: F.D. Por and I. Dor (eds). *Hydrobiology of the Mangal*. The Hague: Dr W. Junk Publishers, 89–109.

Keenan, C.P., P.K.F. Davie, and D.L. Mann. 1998. "A revision of the genus *Scylla* De Haan, 1833 (Crustacea: Decapoda: Brachyura: Portunidae)." *Raffles Bulletin of Zoology* 46(1): 217–245.

Lee, S.Y. 1989. "The importance of sesarmine crabs *Chiromanthes* spp. and inundation frequency on mangrove (*Kandelia candel* (L.) Druce) leaf litter turnover in a Hong Kong tidal shrimp pond." *Journal of Experimental Marine Ecology* 131: 23–43.

MacNae, W. 1968. "A general account of the fauna and flora of mangrove swamps and forests in the Indo-West Pacific region." *Advances in Marine Biology* 6: 73–270.

Mia, M.Y., S. Shokita, K. Kamizato, and M. Kinjo. 1999. "Feeding habits of the grapsid crab, *Helice leachi* Hess, under laboratory conditions." *Bulletin of the Faculty of Science of the University of Ryukyus* No. 68: 31–44.

Micheli, F. 1993. "Feeding ecology of mangrove crabs in north eastern Australia: mangrove litter consumption by *Sesarma messa* and *Sesarma smithi*." *Journal of Experimental Marine Biology and Ecology* 171: 165–186.

Miyake, S. 1983. *Japanese Crustacean Decapods and Stomatopods in Color. Vol. II Brachyura (Crabs).* Tokyo: Hoikusha, 277 pp. (in Japanese).

Nakasone, Y. 1977. "Ecological distribution of macrobenthos in mangrove swamps." In: *Urgent Investigations of Mangrove Forest in Kesaji Bay II.* Okinawa Prefecture: The Board of Education (in Japanese), 9–38.

Nakasone, Y. and M. Agena. 1984. "Role of crabs as the degraders of mangrove litters in the Okinawan mangals." In: S. Ikehara and N. Ikehara (eds). *Ecology and Physiology of the Mangrove Ecosystem.* Okinawa: University of the Ryukyus, 153–167.

Nakasone, Y., S. Limsakul, and K. Tirsrisook. 1985. "Degradation of leaf litter by grapsid crabs and a snail in the mangrove forests of Ao Khung Kraben and Mae Nam Wen, Thailand." In: K. Nozawa (ed.). *Mangrove Estuarine Ecology in Thailand.* Thai–Japanese cooperative research project on mangrove productivity and development, 1983–1984. Kagoshima: Kagoshima University, 21–38.

Nishihira, M. 1983. "Grazing of the mangrove litters by *Terebralia palustris* (Gastropoda: Potamidae) in the Okinawan mangal: Preliminary report." *Galaxea* 2: 45–58.

Nishihira, M., M. Tsuchiya, and H. Kubo. 1988. "Distribution of gastropods and degradation of mangrove leaves by the giant potamodid snail *Terebralia palustris* (L.) in Nakamagawa mangal. Iriomote Island, Okinawa." In: R. Marumo (ed.). *Studies on Dynamics and Conservation of Mangrove Ecosystem.* Research Report of *Environment Science*. Tokyo: The Japanese Ministry of Education, Science and Culture, 48–60.

Odum, W.E. 1971. *Pathways of Energy Flow in a Southern Florida Estuary. Sea Grant Tech. Bull. No. 7.* Miami, Florida: University of Miami Sea Grant Program (Living Resources), 162 pp.

Odum, W.E. and E.J. Heald. 1975. "The detritus-based food web of an estuarine mangrove community." In: L.E. Cronin (ed.). *Estuarine Research 1.* New York: Academic Press, 265–286.

Plaziat, J.C. 1977. "Les Cerithides tropicaux et leur polymorphisme lié à l'écologie littorale des mangroves." *Malacologia* 16: 34–44.

Plaziat, J.C. 1984. "Mollusk distribution in the mangal." In: F.D. Por and I. Dor (eds). *Hydrobiology of the Mangal*. The Hague: Dr W. Junk Publishers, 111–143.

Por, F.D. 1984. "Editor's note on mangal fishes of the world." In: F.D. Por and I. Dor (eds). *Hydrobiology of the Mangal*. The Hague: Dr W. Junk Publishers, 207–209.

Robertson, A.I. 1986. "Leaf-burying crabs: their influence on energy flow and export from mixed mangrove forests (*Rhizophora* spp.) in northeastern Australia." *Journal of Experimental Marine Ecology* 102: 237–248.

Robertson, A.I. 1988. "Decomposition of mangrove leaf litter in tropical Australia." *Journal of Experimental Marine Biology and Ecology* 116: 236–247.

Robertson, A.I. and R.A. Daniel. 1989. "The influence of crabs on litter in processing in high intertidal mangrove forests in tropical Australia." *Oecologia* 78: 191–198.

Sakai, T. 1976. *Crabs of Japan and the Adjacent Seas*. (English text, xxix + 773 pp., figs. 1–379). Tokyo: Kodansha.

Sasekumar, A. 1974. "Distribution of macrofauna on a Malayan mangrove shore." *Journal of Animal Ecology* 43: 1–69.

Shokita, S. 1989. *Macrofaunal Community Structure and Food Chain at the Mangals*. Training course on life history of selected species of flora and fauna in mangrove ecosystem, UNDP/UNESCO Regional Project Ras/86/120. 46 pp.

Shokita, S. 1991. "The environmental characteristics of coral reef areas." In: S. Shokita, K. Kakazu, T. Touma, and A. Tomori (eds); Yamaguchi et al. (Trans.). *Aquaculture in Tropical Areas*. Tokyo: Midori Shobo Co., Ltd, 7–24.

Shokita, S. and S. Nishijima. 1983. "Distribution and standing stock of the macrobenthos in the mangrove area of Funaura in Iriomote Island, Okinawa." In: *Survey Report for the Fishing Ground Development in Iriomote Island*. Okinawa: Okinawa Development Agency, 14–27.

Shokita, S., K. Nozawa, N. Yoshikawa, and S. Limusakul. 1983. "Macrofauna in mangrove areas in Thailand." In: K. Nozawa (ed.). *Mangrove Ecology in Thailand*. Tokyo: The Japanese Ministry of Education, Science and Culture, 33–62.

Shokita, S., T. Ameku, H. Yoshida, and H. Morishima. 1988. "Mangal fishes and their food habits in Okinawa, the Ryukyu Islands." In: *Studies on Dynamics and Conservation of Mangrove Ecosystem*. Research Report on Environmental Science. Tokyo: The Japanese Ministry of Education, Science and Culture, 61–76.

Shokita, S., M. Iriondo, A. Kawakami, and V. Havanont. 1998. "Distribution and abundance of crustaceans in the mangrove swamp of the Okukubi River, Okinawa Island." *Annual Report of the Interdisciplinary Research Institute of Environmental Science* 17: 61–72.

Stephenson, T.A. and A. Stephenson. 1949. "The universal features of zonation between tide marks on rocky coasts." *Journal of Ecology* 37: 289–305.

Twilley, R.R., M. Pozo, V.H. Garcia, V.H. Rivera-Monroy, R. Zambrano, and A. Bodero. 1997. "Litter dynamics in riverine mangrove forests in the Guayas River, Ecuador." *Oecologia* 111: 109–122.

Warner, G.F. 1977. *The Biology of Crabs*. London: Paul Elek (Scientific Books) Ltd, 201 pp.

Photo 5.1 Crustaceans found in Okinawan mangals (I): (A) *Thalassina anomala*; (B) *Mictyris brevidactylus*; (C) ecdysis of *Scylla serrata*; (D) *Uca (Amphiuca) chlorophthalma crassipes*; (E) *Helice leachi*; (F) mating of *Neoepisesarma N lafondi* under laboratory conditions

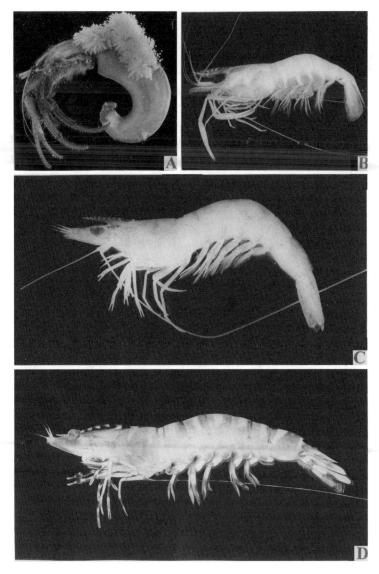

Photo 5.2 Crustaceans found in Okinawan mangals (II): (A) ovigerous *Clibanarius longitarsus*; (B) ovigerous *Macrobrachium scabriculum* living at the boundary between brackish and fresh water; (C) *Metapenaeus moyebi*; (D) *Penaeus monodon*

Photo 5.3 Molluscs found in Okinawan and Thai mangals: (A) *Geloina coaxans*, muddy channel, Iriomote Island; (B) clusters of *Terebralia palustris*, rasping fallen mangrove leaves, Ao Khung Kraben mangal, Chanthaburi Province of eastern Thailand; (C) *Telescopium telescopium* on the mud-flat of Ranong mangal, Thailand; (D) *Rhizophomurex capucinus* preying on the young of *Terebralia palustris*, Ao Khung Kraben mangal; (E) *Cerithidea quadratai* hanging from a dead mangrove branch, Ao Khung Kraben mangal; (F) *Elobium auris-judae* at rest on the trunk of *Rhizophora* sp.; (G) a cluster of *Cerithidea rhizophorarum* at rest on the trunk of *Rhizophora stylosa*, Ishigaki Island mangal, Okinawa

Photo 5.4 Tree-dwelling fauna on Ranong mangal, Thailand: (A) *Aegiceras corniculatum* at the seaward edge of transect 2, Ranong, with sedentary species such as *Melina ephippium*, *Saccostrea cucullata*, and *Balanus amphitrite*; (B) a large colony of *M. ephippium* attached by its byssus to the trunk of *A. corniculatum*; (C) *Balanus amphitrite* attached to the trunk and branches of *Sonneratia alba*; (D) *S. cucullata* on the prop roots of *Rhizophora mucronata* growing on the creek side of transect 3; (E) *Cymia gralata* preying on the *S. cucullata* encrusting the roots of *A. corniculatum*

Photo 5.5 Fish species found in Okinawan mangals (I): (A) *Carcharinus leucus* (also shown here are *Mugil cephalus*, *Platycephalus* sp., and *Acanthopagrus* sp.); (B) *Evenchelys macrurus*; (C) *Mugil cephalus*; (D) *Sphyraena barracuda*; (E) *Apogon amboinensis*; (F) *Caranx sexfasciatus*

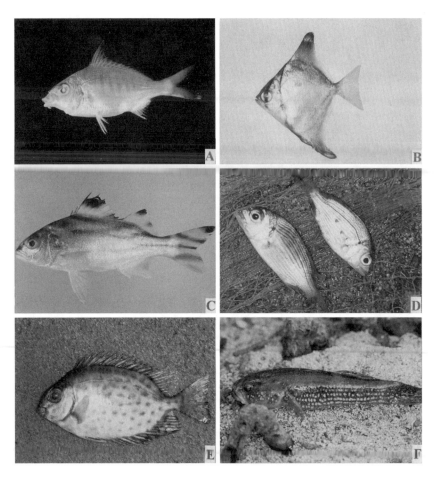

Photo 5.6 Fish species found in Okinawan mangals (II): (A) *Gerres filamentosus*; (B) *Monodactylus argenteus*; (C) *Terapon jarbua*; (D) *Acanthopagrus silvicolus*; (E) *Siganus guttatus*; (F) *Ophiocara poracephala*

6

Effects of mangrove restoration and conservation on the biodiversity and environment in Can Gio District

Phan Nguyen Hong

Introduction

Can Gio is a suburban marine district located south-east of Ho Chi Minh City (formerly Saigon; latitude 10°22'14"–10°40'09"N; longitude 106°46'12"–107°00'59"E) and covers an area of 71,360 ha. It is 35 km long from north to south and 30 km wide from east to west (fig. 6.1). The extent of mangrove forest land accounts for 54.2 per cent of the total natural area of the district (approximately 38,750 ha). During the two Indo-Chinese wars, Can Gio mangroves were used as bases for the resistance, so they were protected; however, during the second Viet Nam war, the mangroves were sprayed with, and devastated by, herbicides. After these many years, the degraded land still remains bare or with sparse bushes.

Since 1978, a major reforestation programme has been undertaken by the Ho Chi Minh City Administration. From 1991, the Government of Viet Nam confirmed and approved Can Gio mangrove forests as environmental protection forests. The restoration and conservation of mangroves has brought about a vast ecological improvement in the environment and rich biodiversity. Can Gio forests have become one of the most beautiful of the rehabilitated mangrove areas of the world. The excellent results of such restoration and conservation work led, in January 2000, to the recognition, by the Bureau of the International Coordinating Council of UNESCO's Man and Biosphere (MAB) Programme, of Can

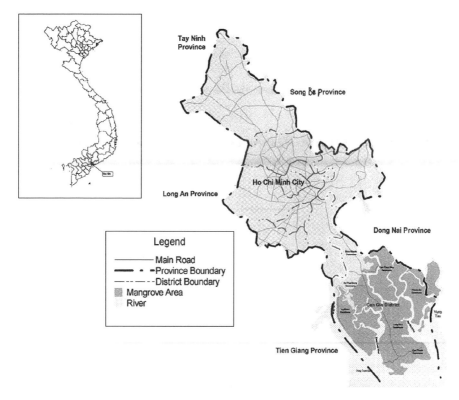

Figure 6.1 Map of the Can Gio District, Ho Chi Minh City

Gio as a mangrove biosphere reserve, the first biosphere reserve internationally acknowledged in Viet Nam.

This chapter presents an analysis of the progress of Can Gio mangroves and of the changes in the natural resources and environment since the inception of the long-term programme of reforestation and conservation.

The natural conditions

Climate

Can Gio is situated in the typical tropical monsoon zone, with two distinctive seasons: the dry season lasts from November to the following April; the rainy season is from mid-May to the end of October.

The temperature is high and stable, the mean monthly temperature being 25.5–29.0°C.

The east monsoon winds blow during the entire dry season (from October to the following April); the south-west monsoon brings the rainy season. The monsoons produce a series of waves with different directions and coastal currents which influence directly Can Gio's coastal zone and lead to erosion and sea-level rise.

The rainy season starts on 15–25 May and ends late in October (25–31). The average annual rainfall is about 1,336 mm, September seeing the highest rainfall (300–400 mm).

The mean annual relative humidity is 80 per cent; however, in the dry season, the daytime air humidity is usually below 60 per cent.

The average rate of evaporation is 3.5–6 mm/day and – in March and April – 7–8 mm/day at most; it is 2.5–5.0 mm/day in the rainy months.

Hydrology

Tides

Can Gio has an irregular semi-diurnal tidal regime of high amplitude (3.3–4.1 m). This great amplitude, coupled with the monsoon rains, creates swift currents that cause erosion of the substratum of the main rivers.

Salinity

The greatest water salinity occurs in the months of March and April: it reaches 19–20 ppt in the northern area and 26–30 ppt near the sea. In the rainy season, from August to October the water salinity in the mangrove areas is only 4–8 ppt. The average monthly salinity is 18 ppt.

Historical development of mangroves in Can Gio

Mangrove status before the chemical war

From 1917 all mangrove forests were periodically exploited by the Water and Forestry Bureau of the French Colonial Government. In 1929, the French colonists enacted an ordinance to conserve the protected mangrove forests in the Quang Xuyen and Can Gio districts (Nam and Quy 2000). Favourable natural conditions and adequate management encouraged the major development of mangroves. The dominant species was, on average, 12–15 m high with a diameter of 20–30 cm.

Cuong (1964) categorized mangrove vegetation in this area into two groups – those growing in salt water and those growing in brackish water.

Salt-water communities

- The pioneer community growing on the new mud-flats of estuaries consisted of *Sonneratia alba* and *Avicennia alba*. The pioneer population of *A. alba* developed along the river-banks.
- The community of *Rhizophora apiculata* and *S. alba* (50 and 20 per cent, respectively) was formed on the mud-flats stabilized by the pioneer community or population. The mangroves in this area were 8–10 m high. The following species could also be found in this community: *Xylocarpus granatum, Kandelia candel, Derris trifoliata,* and *Aegiceras corniculatum*.
- The community of *X. granatum* and *R. apiculata* became established on land inundated to a depth of 2–2.5 m. *R. apiculata* was 15–18 m in height. Other species present were *Ceriops tagal, A. alba,* and *X. granatum*. This is a common type of mangrove community.
- The community of *R. apiculata* and *C. tagal* was distributed on those mud-flats flooded by tides to a depth of 2.5–3 m. *R. apiculata* was dominant here and over 10 m high. *Avicennia officinalis, X. granatum, X. moluccensis,* and *Ceriops decandra* also occurred in this community.
- On those mud-flats inundated by tides to a level of 2.5–3 m *A. officinalis* and *C. decandra* were common. Other species found were *X. moluccensis, C. tagal,* and *Lumnitzera littoralis*.
- The community of *Excoecaria agallocha* and *Phoenix paludosa* was observed only on land flooded by high tides (3.3–4.0 m). Additionally, *X. moluccensis, C. tagal, Acrostichum aureum, Heritiera littoralis,* and *Flagellaria indica* were present.

Brackish-water communities

Brackish-water communities were distributed along the river-banks of brackish-water areas in the north of Can Gio District and along the Dong Nai estuary. There were four communities, classified as follows:
- In areas inundated to 1–1.5 m at high tide, the pioneer species found was *Sonneratia caseolaris,* 8–10 m in height.
- In areas flooded by tides to a depth of 1.5–2 m, the community consisted of *Cryptocoryne ciliata* and *Acanthus ebracteatus,* together with *Nypa fruticans, Cyperus malaccensis,* and *Derris heterophylla*.
- In areas flooded by tides to a level of 2–3 m, the community of *Annona reticulata* and *Flagellaria indica* was found. Other species included *Amoora cucullata, Barringtonia acutangula,* and *Gardenia lucida*. These mangrove trees were 6–8 m tall.
- In the areas inundated to 3–4 m at high tide, a community of *Melastoma polyanthum* and *Dalbergia canadensis* was formed, together with other species such as *Pandanus tectorius, Glochidion littorale, Hibis-*

cus tiliaceus, *Thespesia populnea*, *Clerodendron inerme*, and *Pluchea indica*.

Although no statistical data regarding animals in Can Gio were available, elderly inhabitants recalled that, before the August revolution, many terrestrial animals – such as deer (including sambars), monkeys, pythons, monitor lizards, and tigers – had lived in the mangroves. Sometimes crocodiles could be found in the Dong Tranh and Soai Rap rivers. The names of some channels (such as crocodile channel, wild boar channel, and wolf channel) indicated the previous presence of these animals.

Mangrove status during the second Viet Nam war

In the second Viet Nam war, mangrove forests in Can Gio were one of the bases of resistance in south-east Viet Nam. During this period, they were extensively damaged by bombs and herbicides.

From 1966 to 1970, with the launching of Operation Ranch Hand, the US Air Force sprayed 1,017,515 gallons of herbicides and defoliants (Ross 1975), destroying the mangrove area of Can Gio.

Heavy defoliation not only exterminated the vegetation but also destroyed the fauna and changed the entire ecosystem (Hong and San 1993). A number of terrestrial animals and birds were killed; some migrated to safer areas. It was found that although, after the defoliation, fish, crustaceans, and molluscs benefited from the decomposition of organic matter, their numbers decreased again within three years (Hong and Tri 1986); no mammalian species survived.

Severe erosion was observed on the bare land along the river and canal banks and the area covered by water increased. Coastal erosion increased because of the wide tidal amplitude and the lack of a protective green belt. The comparison of sprayed areas in 1958 and 1989 has shown that the percentage of water surface further increased to 31.2 per cent (Nam and My 1992).

In the low-lying wetlands, after high tide, seawater and rainwater accumulated and served as breeding places for *Anopheles sundaicus* mosquitoes, a vector of malaria.

Investigation of soil characteristics of the bare land sprayed by herbicides showed that the wastelands contain much more Na^+, SO_4^{2-}, Fe^{3+}, and Al^{3+} than the soil of the remaining forests and the pH value is low, especially in deep layers down to 90–160 cm (Tu 1996). Under the action of strong sunlight, high temperature, and stagnant waters in the dry season, the soil was slowly oxidized and became acid sulphate soil.

On the land flooded only at spring tides, *Phoenix paludosa* and *Acrostichum aureum* developed. Stunted specimens of *Ceriops tagal*, *Excoecaria agallocha*, and *Lumnitzera racemosa* or *Avicennia lanata* regenerated on land flooded at high tide. On the river and canal mud-flats, *A. alba* recolonized the area, as a result of water-borne seeds from other, unimpaired mangrove areas.

Mangrove status after the liberation

After the war (1975), in the herbicide-sprayed mangrove areas the following species could no longer be found: *Rhizophora apiculata, R. mucronata, Bruguiera cylindrica, B. gymnorhiza, B. sexangula*. A regenerated vegetation cover was formed, consisting mainly of *A. officinalis* and *Ceriops decandra* less than 4–6 m in height and growing on the area flooded at mid-tide. *A. aureum* and *P. padulosa* could be found on land flooded at high tide. These species rapidly regenerated and covered the barren land with a dense vegetation; however, these regenerated forests were destroyed for use as firewood by the local inhabitants. To deal with the food shortage, from 1975 to 1978 the Ho Chi Minh authorities established a number of agricultural enterprises to cultivate many kinds of food plants on the bare land, but all such efforts ended in failure because of the shortage of fresh water and the acid sulphate condition of the soil (Hong 1991).

As a step towards the restoration policy to form a green belt around the city, the Ho Chi Minh City People's Committee signed Resolution No. 165/QD-UB on 7 August 1978, the aim of which was to establish a mangrove reforestation area in Can Gio. The Duyen Hai Forestry Enterprise was established and delegated to carry out this important task (Hong 1996). From 1978 to 1997, 20,638 ha were under reforestation, using *Rhizophora apiculata* as the main species (Tuan 1998).

Although the achievement of Ho Chi Minh City in reforestation has been encouraging, it has been difficult to protect the newly forested areas because of economic pressure by the local population.

The development of mangrove reforestation in Can Gio can be divided into two periods, as follows:
- *In the 1980s*, many factors caused the degradation of replanted mangroves, such as the imbalance between demand and supply of timber for scaffolding poles for house construction and for firewood, lack of facilities and means of transportation for the forest guards, and the poor educational facilities (Cuong 1994). In particular, the lucrative benefits from shrimp export encouraged local farmers and some agencies in Ho Chi Minh City to clear tracts of mangroves for shrimp farming. After 3 years of shrimp rearing, the mangrove soil and water deteriorated, shrimp yields fell sharply, and numerous shrimp-ponds became useless.
- *From 1991 to the present*, with the government's reform policy, the

Ministerial Council issued Decision No. 173 CT dated 29 May 1991 to confirm and approve Can Gio Mangrove Forest as the Environmental Protection Forest of Ho Chi Minh City.

In 1991, the Duyen Hai Forestry Enterprise was upgraded to the Management Board for Environmental Protection Forests (MBEPF) of Ho Chi Minh City with personnel reinforcement. The city has put into practice the policy of land and forest allocation to households; as a result, the destruction of forests has greatly decreased.

- *At present*, Can Gio forests are divided into 24 forestry units with clearly marked natural boundaries; land maps and exploitation logbooks are maintained to monitor any changes over time (table 6.1, fig. 6.2).

Table 6.1 Can Gio mangroves in 24 forestry units

Forestry units	Rehabilitated forest (ha)	Natural forest (ha)	Other lands (ha)[a]	Total area (ha)
1	643.50	293.30	292.40	1.229.20
2	1.162.98	317.26	451.19	1.931.43
3	549.90	462.70	153.60	1.166.20
4	1.124.10	334.49	223.60	1.682.19
5	719.07	186.84	228.18	1.134.09
6	940.49	349.91	366.28	1.656.68
7	496.40	231.20	444.40	1.172.00
8	933.52	180.38	339.84	1.453.74
9	858.80	407.30	179.50	1.445.60
10	1.126.42	329.76	141.02	1.597.20
11	656.60	288.90	421.90	1.367.40
12	742.20	135.20	458.10	1.335.50
13	879.50	201.00	801.90	1.882.40
14	658.90	280.50	538.90	1.478.30
15	866.00	401.07	673.77	1.940.84
16	642.10	234.20	721.90	1.598.20
17	1.146.71	378.19	690.52	2.215.42
18	651.80	284.30	744.00	1.680.10
19	479.70	295.20	471.90	1.246.80
20	540.05	478.30	891.65	1.910.00
21	641.10	169.40	793.70	1.604.20
22	265.10	236.10	231.25	732.45
23	2.473.50	73.90	200.90	2.748.30
24	1.438.10	475.50	626.90	2.540.50
Total	20.636.54	7.024.90	11.087.30	38.748.74

Source: Tuan (1998).

a. Other lands include salt pans, shrimp ponds, bare land with *Sesuvium portulacastrum* and residential land of guard households in the forest land.

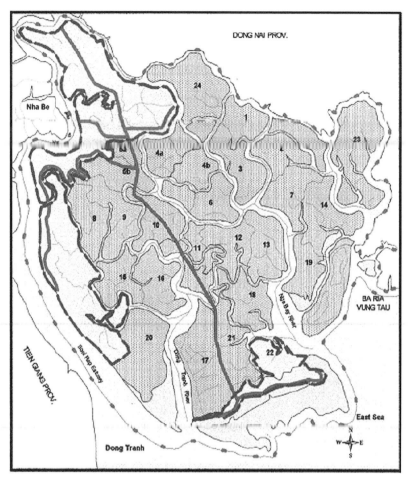

Figure 6.2 Map of Can Gio mangrove forest and forestry units. Source: Management Board for Environmental Protection of Forests

Most guardians include workers from the MBEPF and from agroforestry enterprises, as well as employees of the forestry agencies and householders. Areas of different parts of the forest are allocated to householders for protection according to the 30-year contracts signed with MBEPF; 10,850 ha are allocated to 208 householders.

Effects of mangrove reforestation

Natural resources

Flora

Compared with the flora investigated by Cuong (1964) before the wartime herbicide spraying, although the current natural mangrove flora is similar, the individual numbers and distribution of the various species are not. Our investigation has listed 72 species, of which 30 are true mangroves and 42 associate mangrove species (appendix 6.1). Besides the mangrove flora, 95 species belonging to 42 families of inland plants dispersed by humans and animals have been found. These species form diverse communities of different biotopes.

Vegetation characteristics

At present, the mangroves in Can Gio are more diverse in community structure than were those before the wars. An explanation for this is that, in Can Gio mangroves, valuable species mixed with naturally regenerated ones have been rehabilitated. Some are found naturally regenerating on abandoned shrimp-ponds. Some of the main species in high-saline and in brackish-water communities are as follows.

1. Communities in areas of high salt concentrations:
    - The pioneer community of *Sonneratia alba/Avicennia alba* can be found on the coastal and estuarine newly accreted lands daily flooded by the tides. The salinity here is high.
    - The pioneer population of *A. alba* is present on the newly formed mud-flats along the rivers and creeks.
    - The *Rhizophora apiculata/Avicennia alba* community is seen between the replanted *R. apiculata* and the population of *A. alba* regenerating naturally on deep muddy soil.
    - The *R. apiculata/A. officinalis* community is distributed on firm mud-flats flooded by normal high tides. These two species are found regenerating naturally in groups.
    - The community of *R. apiculata/Lumnitzera racemosa/Ceriops decandra* is formed on firm mud-flats flooded by spring tides. Only *R.*

apiculata was planted between 1981 and 1985; it was later chopped down, leaving behind it bare areas and naturally regenerated shrubs.
- The community of *R. mucronata/Phoenix paludosa/Lumnitzera racemosa* grows on firm mud-flats flooded by spring tides. *Phoenix* was burnt down to spare land for *Rhizophora* planting; the species subsequently regenerated vigorously from its own left-over stumps in the rainy season.
- The community of *Ceriops tagal/C. decandra* can be seen on sandy mud-flats flooded by spring tides. The dominant species is *C. tagal*; *R. apiculata* and *L. racemosa* are sparsely distributed.
- The community of *Phoenix paludosa/Acrostichum aureum* is on firm, sandy, clay flats flooded only by spring tides. The associate species are *L. racemosa*, *Excoecaria agallocha*, and *Pluchea indica*.
- The population of *Excoecaria agallocha* is formed on firm sandy/ loam flats.
- The *Avicennia lanata* population regenerates on the high land of sandy clay or on the substratum of abandoned salt pans.
- The population of *Sesuvium portulacastrum* colonizes wastelands of firm, sandy clay or former salt-pans and shrimp-ponds.

2. Brackish-water communities

Brackish-water communities at present do not differ greatly from those found prior to the wars. Some main communities are:
- Pioneer community of *Sonneratia caseolaris/Avicennia alba*;
- Pure *Nypa fruticans* populations;
- *Nypa fruticans/Acanthus ilicifolius/Cryptocorine ciliata* community;
- *Sonneratia caseolaris/Nypa fruticans* community;
- *Annona glabra/Flagellaria indica* community.

Plankton

The analysis of 52 water samples of plankton from different points in the north–south and east–west directions in Can Gio by Viet (1993) has revealed the presence of 63 phytoplankton species and 19 zooplankton species. Zooplankton had increased considerably in the samples collected in June 1993 (100,000–1,000,000 cells/m^3) compared with that in June 1990 (67,000–6,025,000 cells/m^3).

Benthos

Mangrove rehabilitation has changed the properties of the soil owing to the added sediments formed by litter fall and large quantities of fine and fibrous root matter. The mud contains the highest concentration of organic carbon and nitrogen (Alongi and Sasekumar 1992); this has directly influenced the distribution of the benthos. On the other hand, the benthos plays an important role in the decomposition of litter fall. Preliminary surveys have enabled 118 species of invertebrate macrobenthos to be listed (appendix 6.2). Polychaetes were represented by 32 species of eighteen families; molluscs comprised 15 species of gastropods belonging to eight families and bivalves of 17 species from seven families. Crustaceans were represented by Macrura (28 species belonging to seven families) and Brachyura (25 species of four families) (Anon. 1998; Hong 1996; Mien et al. 1992; appendix 6.2).

Fishes

Fishes use mangrove waters as breeding and/or nursery grounds or as their permanent habitat (Aksornkoae 1993). Of the 133 species found belonging to 40 families (appendix 6.3), the majority are true estuarine species, the remainder being eurythermal and euryhaline. Most species live in shallow waters and often enter estuaries to feed (Hong 1996; Lai 1989; Mien et al. 1992; Tang 1984). Recently, numbers of *Mugil affinis* have increased rapidly, especially in rivers draining mangroves containing abundant *Avicennia alba* and *A. officinalis*; *Lates calcarifer* is concentrated in shrimp-ponds. *Periophthalmus schlosseri*, previously rare, has now become fairly common, particularly at mud-flats near salty estuaries.

Amphibians and reptiles

Since the local people have reforested and excavated shrimp-ponds in mangrove areas, there has been a great increase in amphibians and reptiles, which burrow into the embankments of the ponds and search for prey and other food at night (appendix 6.4).

Preliminary surveys have revealed nine species of amphibians belonging to four families (Dat 1997; Hong 1996). Tree frogs (*Racophorus leucomystax*) are commonly found, their loud croak often being heard in the forests. They breed in the rainy season, leaving a foamy secretion on the trees as a result of mass copulation. The cricket frog (*Rana limnocharis*) is mainly seen in rice fields but also occurs in the brackish waters of the mangroves. The common toad (*Bufo melanostictus*) is often found in guardhouses where mosquitoes abound.

Of the 30 species of 14 families of reptiles found in Can Gio, eight species belong to the family Colubridae and three are Cheloniidae; the remainder are found only in large rivers and on the seashore. The king cobra (*Ophiophagus hannah*) is distributed throughout the natural and replanted forests, while *Naja naja* is widely distributed in villagers' gardens. A number of aquatic snakes, such as *Achrochordus javanicus* and *Lapenius hardwickii*, and marine turtles, such as *Chelonia mydas* and *Leidochelys olivacea*, are only caught along the seashore by fishing nets. In 1983, some workers at the enterprise shot a crocodile, but since then no other crocodiles have been seen. Only 30 crocodiles (*Crocodylus po-*

rosus) have been reared at the Can Gio Forestry Park. Since the 1990s, as a result of the effective protection of forests, many birds have arrived and, consequently, a number of reptile predators on birds' eggs or chicks (such as monitor lizards, boas, and snakes) are on the increase.

Birds

From 1975 to 1980, birds were very rare in Can Gio mangrove areas, especially on high ground flooded only at high tide. Bird droppings were rarely seen on the ground or under the trees. Local people have reported only some kingfishers occasionally perching on *Avicennia* branches, or several little egrets (*Egretta garzetta*) or lesser sand plovers searching for food on the mud-flats far from villages.

Since the 1980s, more and more birds have come back to the area. Beside waterfowl, many flocks of migratory birds from the north (such as *Charadrius alexandrinus*, *Pluvialis squatarola*, *Pluvialis fulva*, and especially *Tringa glareola*, *T. erythropus*, and *Himantopus himantopus*) have been seen.

Some water-bird species (namely *Egretta garzetta*, *Buteroides striatus*, and *Bubulcus ibis*) build their nests from April to the following February in *Phoenix paludosa* stumps and *Avicennia* trees, far from residential areas.

Along the mud-flats of the rivers and creeks, which are periodically inundated, a number of waders can be seen either feeding or roosting on the banks (Hussain and Acharya 1994). About 130 species of 44 families are known to occur in Can Gio: of these, 49 species are waterfowl, including 22 migratory species. The others are common in the delta and are seen in various habitats (Dat 1997; Hong 1996; appendix 6.5).

Birds play an important role in the ecosystem, contributing actively to fertilization of the soil, to enrichment of the ecosystem, and to the production of available food in the mangrove areas.

Mammals

Nineteen species of mammals belonging to thirteen families and an additional five species of chiropterans have been recorded (Dat 1997; appendix 6.6). Among the terrestrial mammals, after mangrove restoration the numerically most abundant species is the macaque monkey (*Macaca fascicularis*). Since 1992 they have multiplied very fast, particularly in the Forestry Park. Daily, at ebb tide, they go to the mud-flats to catch small crabs and mud skippers. Recently, they have been fed by the Forestry Park's staff and tourists and have become increasingly used to the presence of humans. Rodents multiply rapidly, particularly species of the family Muridae, owing to the increase in the human population. Groups of wild boars (*Sus scrofa*) are fairly common in the patches of *Phoenix pal-*

Table 6.2 Rare animals in Can Gio

Scientific name	Status[a]	Scientific name	Status
Fish		*Lepidocheilus olivacea*	V
Elops saurus	R	*Crocodylus porosus*	E
Megalops cyprinoides	R	**Birds**	
Albula vulpes	R	*Pelecanus philippensis*	R
Chanos chanos	T	*Mycteria leucocephala*	R
Nematolosa nasus	E	*Mycteria cinerea*	V
Reptiles		*Leptoptilos javanicus*	R
Gekko gecko	T	*Tringa guttifer*	R
Varanus salvator	V	*Pica pica*	E
Python molorus	V	**Mammals**	
P. reticulatus	V	*Lutva lutva*	T
Bugaris fasciculatus	T	*Aonyx cinerea*	T
Ophiophagus hannals	E	*Felix viverrina*	R
Chelonia mydas	E		
Erethmochelis imbricata	E		

Source: Ministry of Sciences, Technology, and Environment (1992).
a. E, Endangered; V, vulnerable; R, rare; T, threatened.

udosa. Local people report sighting the fishing cat (*Felis viverrina*) in the natural *Avicennia* forests; this species is well adapted to the tidal areas and feeds on fish, birds, bandicoot, oysters, and other small animals (Hussain and Acharya 1994).

Generally speaking, the wide diversity of fauna in the mangroves is attributable to abundant food resources and a wide range of microhabitats. Chapman (1975) recognized six types of habitats: soil surface, soil interior, tide pools, small channels or creeks, fallen logs, and the tree canopy; all types have been recorded in the Can Gio mangrove forests.

In the Mekong coastal area, many animal species have become endangered or threatened as a result of mangrove habitat degradation; however, these species in Can Gio have increased in number (table 6.2).

The environment

The rehabilitation of mangrove forests has resulted in certain changes in the environment and in the ecological processes. The organic debris produced by the mangrove vegetation, together with the shelter it provides, and local environmental conditions promote the enrichment of the food chains and spawning and nursery grounds for many vertebrates and fish. Hussain and Acharya (1994) noted that low-oxygen mangrove soils play an important role in denitrification and removal of toxins from water and sediment.

Previously, the coastal area and river-banks of Can Gio suffered from extensive erosion during the north-east monsoon season. After some years of rehabilitation, mud-flats formed along those river-banks subject to erosion by wave and tidal action. Early in November 1997, an unusually severe storm hit the coastal zone of Southern Viet Nam, causing serious damage to many coastal provinces; however, the mangrove belt at Can Gio was effective in mitigating storm damage.

The rehabilitation of mangrove forests has induced changes in the physical and chemical properties of the soil of the area. The substratum has gradually been transformed into loam and consequently the pH value has increased, indicating a reduction in soil acidity (table 6.3).

Owing to the establishment of pioneer species such as *Avicennia alba* and *Sonneratia ovata*, erosion of river-banks has been reduced and accretion has greatly increased, creating large, tidal, sandy, mud-flats, which lend themselves to blood ark-shell (*Arca* spp.) and clam farming.

Mangrove rehabilitation, combined with controlled water discharge from the Tri An and Dau Tieng reservoirs, has contributed to the reduction of salt intrusion in the agricultural production areas of Can Gio and neighbouring districts, in both the rainy and the dry seasons. Furthermore, surface-water pollution caused by waste water flowing from Ho Chi Minh City through the vast mangrove areas has been on the decrease. A survey by the Center for Environment Protection has shown that organic pollution in the district of Nha Be is much more serious than that in the mangrove area to the south of Can Gio. The physicochemical indices and the concentration of harmful micro-organisms are lower in the mangrove areas than in those adjacent.

Mangrove restoration also offers good opportunities for eco-tourism, biological research, and conservation education. Recently, Action for Mangrove Reforestation, Japan (ACTMANG) has supported Ho Chi Minh City in constructing a model for a Human Ecology Park in the Can Gio mangrove area for Vietnamese and foreign students.

Conclusions

Recognizing the importance of the mangrove ecosystem in Can Gio, the local and central authorities have agreed to provide major investment capital for mangrove restoration.

The rehabilitation of mangroves has ensured conservation of soil and water, an increase in biodiversity, and the development of eco-tourism with improvement of local living standards. However, this fragile ecosys-

Table 6.3 Physicochemical characteristics of mangrove soils after reforestation

Sample		Layer (depth)	pH H$_2$O	Salinity (%)	Cl$^-$ (%)	SO$_4^{2-}$ (%)	Al^{3+} (ppm)	Total (%)					Structural composition (%)		
No.	Character							N	P$_2$O$_5$	Fe$_2$O$_3$	Ca	Mg	Clay	Loam	Sand
1	Accreted mud		5.85	2.69	1.33	0.79	3.90	0.201	0.100	7.84	0.17	0.71	51.6	32.8	15.6
2	*Avicennia alba* Com.	Surface	6.73	1.62	0.86	0.19	0.28	0.164	0.095	6.23	0.19	0.67	56.4	38.1	5.5
		50 cm	6.78	1.76	0.90	0.24	0.36	0.187	0.096	6.63	0.14	0.70	57.1	32.9	10.1
3	*Rhizophora apiculata* Com.	Surface	6.58	1.84	0.92	0.38	0.56	0.211	0.105	6.06	0.20	0.58	44.7	38.5	16.8
		50 cm	5.16	2.78	1.55	1.04	5.90	0.268	0.057	4.58	0.26	0.45	50.7	32.4	16.9
4	*Phoenix paludosa* Com.	Surface	7.21	0.75	0.35	0.06	0.00	0.192	0.112	5.58	0.14	0.64	46.8	45.9	7.3
		50 cm	7.23	1.39	0.71	0.15	0.01	0.147	0.081	5.60	0.14	0.55	51.6	40.0	8.4
		100 cm	7.12	1.81	0.98	0.19	0.04	0.138	0.080	5.68	0.19	0.55	54.6	32.5	12.9

Source: Lai (1997).

tem, which can be rehabilitated only by the expenditure of much time and labour, should be borne in mind when the development of the economy of the district and the preparation of a tourism plan in the new century are being considered, or it will again be endangered.

In November, 1998, a workshop on "Can Gio mangrove biosphere reserve preparation and biosphere reserve networking in the southeast Asian Region" was held in Ho Chi Minh City, Viet Nam by MAB Viet Nam and MERD with the participation of some experts in mangroves and Biosphere Reserve (Anon. 1998).

At present, the Can Gio mangrove area is recognized as a mangrove biosphere reserve. It is a unique and complete ecosystem, of value not only for its multiple natural resources but also for its outstanding scientific and educational interest.

It is clear that close cooperation and collaboration between domestic agencies and international organizations for the planning and implementation of all activities involving the Can Gio Mangrove Biosphere Reserve is imperative. All possible measures should be taken to develop and implement an effective management plan. In addition, continued research into the functioning and requirements of the mangrove ecosystem in Can Gio is essential to create a sound, scientific basis for a management plan regarding mangrove resources throughout the country.

Appendices

Appendix 6.1 Mangrove flora in Can Gio Mangrove Biosphere Reserve

True mangroves	Associate mangroves	
Acanthaceae	Amaryllidaceae	Loranthaceae
1 *Acanthus ebracteatus*	1 *Crinum asiaticum*	31 *Viscum ovalifolium*
2 *A. ilicifolius*	Annonaceae	Malvaceae
Aizoaceae	2 *Annona glabra*	32 *Hibiscus tiliaceus*
3 *Sesuvium portulacastrum*	Apocynaceae	33 *Thespesia populnea*
	3 *Cerbera manghas*	Myrtaceae
Araceae	4 *Cerbera odollam*	34 *Melaleuca cajeputi*
4 *Cryptocoryne ciliata*	Araceae	Papilionaceae
Avicenniaceae	5 *Lasia spinosa*	35 *Derris trifoliata*
5 *Avicennia alba*	Asclepiadaceae	36 *Pongamia pinnata*
6 *A. lanata*	6 *Finlaysonia obovata*	Salvadoraceae
7 *A. officinalis*	7 *Gymnanthera nitida*	37 *Azima sarmentosa*
Bignoniaceae	8 *Sarcolobus globosus*	Sapindaceae
8 *Dolichandrone spathacea*	Asteraceae (Compositae)	38 *Allophylus glaber*
		Styracaceae
Combretaceae	9 *Pluchea indica*	39 *Styrax agrestis*
9 *Lumnitzera littorea*	10 *Tridax procumbens*	Verbenaceae
10 *Lumnitzera racemosa*	11 *Wedelia biflora*	40 *Clerodendron inerme*
Euphorbiaceae	Boraginaceae	41 *Premna integrifolia*
11 *Excoecaria agallocha*	12 *Cordia cochinchinensis*	Vitaceae
Meliaceae		42 *Cayratia trifolia*
12 *Xylocarpus granatum*	Caesalpiniaceae	
13 *X. moluccensis*	13 *Dalbergia candenatensis*	
Myrsinaceae		
14 *Aegiceras corniculatum*	14 *Intsia bijuga*	
Palmae	Chenopodiaceae	
15 *Nypa fruticans*	15 *Suaeda maritima*	
16 *Phoenix paludosa*	Combretaceae	
Pteridaceae (fern)	16 *Combretum quadrangulare*	
17 *Acrostichum aureum*		
Rhizophoraceae	17 *Terminalia catappa*	
18 *Bruguiera gymnorrhiza*	Convolvulaceae	
19 *B. parviflora*	18 *Ipomoea pes-caprae*	
20 *B. sexangula*	Cyperaceae	
21 *Ceriops decandra*	19 *Cyperus elatus*	
22 *C. tagal*	20 *C. malaccensis*	
23 *Kandelia candel*	21 *C. tagitiformis*	
24 *Rhizophora apiculata*	22 *C. soloniferus*	
25 *R. mucronata*	23 *Fimbristylis littoralis*	
Rubiaceae	24 *F. ferruginea*	
26 *Scyphiphora hydrophyllacea*	Flagellariaceae	
	25 *Flagellaria indica*	
Sonneratiaceae	Gramineae	
27 *Sonneratia alba*	26 *Cynodon dactylon*	
28 *S. caseolaris*	27 *Phragmites karka*	
29 *S. ovata*	28 *Sporobolus virginicus*	
Sterculiaceae	Lecythidaceae	
30 *Heritiera littoralis*	29 *Barringtonia acutangula*	
	30 *B. asiatica*	

Sources: Nam and My (1992); Hong (1991, 1997b).

Appendix 6.2 Zoobenthos in Can Gio Mangrove Biosphere Reserve

POLYCHAETA
Nereidae
1 *Namalycatis aibiuma*
2 *Dendronereis aestuarina*
3 *Tylorhynchus heterochaetus*
4 *Neanthes succinea*
5 *Ceratonereis mirabilis*
Eunicidae
6 *Diopatra neapolitana*
Phyllodocidae
7 *Eteone (Mysta) ornata*
Glyceridae
8 *Glycinde nipponica*
9 *G. armigera*
Nephthydidae
10 *Nephthys polybranchia*
11 *N. gravieri*
12 *N. californiensis*
13 *N. oligobranchia*
Spionidae
14 *Prionospio japonicus*
Owenidae
15 *Owenia fusiformis*
Ampharetidae
16 *Amphreta longipaleolata*
Terebellidae
17 *Maldane sarsi*
18 *Terebellides stroemi*
Ariciidae
19 *Scoloplos armiger*
Sabellidae
20 *Bispira polymorpha*
21 *Sabellaria cementarium*
22 *Sternaspis sculata*
Chloraemidae
23 *Stylaroides plumosus*
Aphroditidae
24 *Aphrodita australis*
25 *Sigalion papillosum*
Amphinomidae
26 *Chloeia flava*
Euphrosynidae
27 *Euphrosyne hortensis*
Cirratulidae
28 *Chaetozone setosa*
29 *Cossura longicirrata*
Maldanidae
30 *Maldane sarsi*
31 *Asychis gotoi*

Trichobanchiidae
32 *Terebellides stroemi*
SIPUNCULIDA
Sipunculidae
33 *Phascolosoma arcuatum*
CRUSTACEA
MACRURA
Balanidae
34 *Balanus amphritrite*
Alpheidae
35 *Alpheus malabaricus*
36 *A. rapacida*
Palaemonidae
37 *Macrobrachium nipponense*
38 *M. lanchesteri*
39 *M. mirabilis*
40 *M. equidens*
41 *M. rosenbergii*
42 *Exopalaemon carinicauda*
43 *Exopalaemon styliferus*
Penaeidae
44 *Penaeus indicus*
45 *Penaeus merguiensis*
46 *P. semisulcatus*
47 *P. monodon*
48 *P. japonicus*
49 *Metapenaeus ensis*
50 *M. affinis*
51 *M. tenuipes*
52 *M. lysianassa*
53 *M. brevicornis*
54 *Parapenaeopsis hardwickii*
55 *P. cultrirostris*
56 *P. hungerfordi*
Atyidae
57 *Caridina weberi sumatresis*
58 *C. nilotica*
Palinuridae
59 *Panulirus ornatus*
Squillidae
60 *Squilla scorpioides*
61 *S. choprai*
BRACHYURA
Grapsidae
62 *Sesarma bidens*
63 *S. plicata*
64 *Metaplax longipes*
65 *M. elegans*
66 *Helice tridens tientsiensis*

Appendix 6.2 (cont.)

Ocypodidae
67 *Ocypode ceratophthalma*
68 *Dotilla wichmani*
69 *Ilyoplax orientalis*
70 *I. pingi*
71 *I. serrata*
72 *Uca lacteus*
73 *U. arcuata*
74 *U. demani typhoni*
75 *U. bellator*
76 *U. polita*
77 *U. vocans*
78 *U. forcipata*
79 *U. clorophthalmus*
80 *Scopimera tuberculata*
81 *S. intermedia*
82 *Timethypocoelis ceratophora*
Leucosiidae
83 *Phylira olivacea*
Portunidae
84 *Scylla serrata*
85 *Portunus trituberculatus*
86 *P. pelagicus*
MOLLUSCA
GASTROPODA
Neritidae
87 *Nerita albicilla*
88 *Neritina violacea*
Littorinidae
89 *Littorina brevicula*
90 *L. scabra*
Assimineidae
91 *Assiminea brevicula*
Melanidae
92 *Melanoides tuberculatus*
Potamididae
93 *Cerithidea cingulata*

94 *Bittium craticulatum*
95 *Terebralia sulcata*
Cerithiidae
96 *Cerithium articulatum*
Muricidae
97 *Chicorius ramosus*
Nassariidae
98 *Nassarius arcularius*
99 *N. distortus*
100 *Ellobium aurisjudea*
101 *Cassidula mustelina*
BIVALVIA
Arcidae
102 *Area subcrenata*
103 *A. granosa*
104 *A. ventricosa*
105 *A. antiquata*
Anomiidae
106 *Placuna placenta*
Ostreidae
107 *Ostrea cucullata*
108 *O. modax*
109 *O. rivularis*
Veneridae
110 *Meretrix meretrix*
111 *M. lusoria*
112 *M. lyrata*
Solenidae
113 *Solen gouldi*
114 *S. arcuatus*
115 *Sinonovacula constricta*
Corbiculidae
116 *Cyrena sumatraensis*
Teredinidae
117 *Teredo navalis*
118 *Bankia sauli*

Sources: Mien et al. (1992), Hong (1996), Lai (1997), Nhuong (2000).

Appendix 6.3 Fish in Can Gio Mangrove Biosphere Reserve[a]

OSTEICHTHYS
Elopidae
1 *Elops saurus* R+
Megalopidae
2 *Megalops cyprinoides* R+
Albulidae
3 *Albula vulpes* R+
Chanidae
4 *Chanos chanos* T+
Clupeidae
5 *Nematolosa nasus* E+
6 *Ilisha elongata*
7 *I. indica*
8 *I. dussumieri*
9 *I. pristigatroides*
10 *Macrura kele*
11 *M. sinensis*
Engraulididae
12 *Stolephorus commersonii*
13 *S. tri*
14 *S. heterolobus*
15 *Lycothrissa crocodilus*
16 *Coilia grayii richardson*
17 *C. mystus*
18 *C. dussumieri*
19 *C. macrognathus*
20 *Harpodon nehereus*
Muraenesocidae
21 *Muraenesox cinereus*
22 *M. talabon*
23 *M. talabonoides*
Ophichthyidae
24 *Pisoodonophis boro*
25 *P. cancrivorus*
Synbranchidae
26 *Macrotrema caligans*
Plotosidae
27 *Plotosus anguillaris*
28 *P. caninus*
Schilbeidae
29 *Pangasius krempfi*
30 *P. polyuranodon*
Ariidae
31 *Arius maculatus*
32 *A. sciurus*
33 *A. maculatus*
34 *A. truncatus*
35 *A. caelatus*
36 *A. sagor*

Cyprinodontidae
37 *Aplocheilus panchax*
Belonidae
38 *Tylosurus strongylurus*
Harpodontidae
39 *Tylosurus leiurus*
40 *Xenentodon canciloides*
41 *Ablennes annastomella*
Hemirhamphidae
42 *Hyporhamphus unifasciatus*
43 *H. far*
Sygnathidae
44 *Sygnathus diarong*
Sphyraenoidae
45 *Sphyraena jello*
46 *S. langsar*
Mugilidae
47 *Mugil cephalus*
48 *M. dussumieri*
49 *M. strongycephalus*
50 *M. affinis*
51 *Liza tade*
52 *L. vaigiensis*
Polynemidae
53 *Eleutheronema tetradactylus*
54 *Polynemus plebejus*
55 *P. sextarius*
56 *P. paradiseus*
Centropomidae
57 *Lates calcarifer*
58 *Psammoperca waigiensis*
59 *Ambassis gymnocephalus*
60 *A. commersoni*
61 *A. urotaenia*
Serranidae
62 *Epinephelus areolatus*
63 *E. awoara*
Theraponidae
64 *Therapon puta*
65 *T. theraps*
66 *T. jarbua*
67 *Pelates quadrilineatus*
Leiognathidae
68 *Leiognathus splendens*
69 *L. ruconius*
70 *L. lineolatus*
71 *L. insidiator*
72 *L. fasciatus*
73 *L. equulus*

Appendix 6.3 (cont.)

74 *L. brevirostris*	104 *G. sparsipapillus*
75 *L. daura*	105 *Brachygobius sua*
76 *L. bindus*	Periophthalmidae
77 *L. elongatus*	106 *Periophthalmus schlosseri*
Sparidae	107 *Boleophthalmus boddarti*
78 *Crenidens caisphorus*	Apocrypteidae
Sciaenidae	108 *Pseudapocryptes lanceolatus*
79 *Pseudosciaena soldad*	109 *Parapocryptes serperaster*
Scatophagidae	Gobioididae
80 *Scatophagus argus*	110 *Taenioides gracilis*
Pomacentridae	111 *T. nigrimarginatus*
81 *Abudefduf bengalensis*	112 *Trypauchen vagina*
82 *A. aureus*	Psettodidae
83 *A. bankieri*	113 *Psettodes erumeri*
84 *A. melas*	Soleidae
85 *A. cyaneus*	114 *Zebrias zebra*
86 *A. coelestinus*	115 *Synaptura orientalis*
87 *A. glaucus*	Cynoglossidae
88 *A. curacao*	116 *Paralagusia bilineata*
89 *A. saxatilis*	117 *Cynoglossus bilineatus*
Siganidae	118 *C. macrolepidosta*
90 *Siganus fuscescens*	119 *C. lingua*
Toxotidae	120 *C. cynoglossus*
91 *Toxotes chatareus*	121 *C. puncticeps*
Eleotridae	122 *C. abbreviatus*
92 *Eleotris fusca*	123 *C. gracilis*
93 *Butis butis*	124 *C. xiphoideus*
94 *Oxyeleotris siamensis*	125 *C. wandersi*
Gobiidae	Lagocephalidae
95 *Pohonogobius planifrons*	126 *Torquigener oblongus*
96 *Pseudogobioptis oligastis*	127 *Gastrophysus scleratus*
97 *Oxyurichthys microlepis*	Tetraodontidae
98 *Stigmatogobius sadanundio*	128 *Monotretus cutcutia*
99 *S. javanicus*	129 *Chelonodon fluviatilis*
100 *Acentrogobius viridipunctatus*	130 *Tetraodon palembagensia*
101 *A. atripinnatus*	131 *T. lorteli*
102 *Auloparia jenetae*	132 *T. leiurus*
103 *Glossogobius giuris*	133 *Tetraodon* sp.

Sources: Mien et al. (1992); Hong, Nhuong, and Tuan (1996); Lai (1997).
a. +, Species listed in the *Red Data Book of Vietnam* (Ministry of Sciences, Technology, and Environment 1992); E, endangered; R, rare; T, threatened.

Appendix 6.4 Amphibians and reptiles in Can Gio Mangrove Biosphere Reserve[a]

AMPHIBIA
Ranidae
1 *Rana rugulosa*
2 *R. cancrivora*
3 *R. limnocharis*
4 *Ooedozyga lima*
Bufonidae
5 *Bufo melanostictus*
Rhacophoridae
6 *Rhacophorus leucomystax*
Microhylidae
7 *Kaloula pulchra*
8 *Microhyla pulchra*
9 *M. ornata*
REPTILIA
Gekkonidae
10 *Gekko gecko* T
11 *Hemidactylus frenatus*
12 *H. bowringi*
Agamidae
13 *Calotes versicolor*
14 *Leiolepis belliana*
Lacertilidae
15 *Takydromus* sp.
Scincidae
16 *Mabouya multifasciata*
17 *M. macularia*
Varanidae
18 *Varanus salvator* V
Boidae

19 *Python molurus* V
20 *P. reticulatus* V
Uropeltidae
21 *Cylindrophis rufus*
Colubridae
22 *Homalopsis buccata*
23 *Cyclophiops maijor*
24 *Xenochrophis piscator*
25 *Cerberus rhynchops*
26 *Elaphe radiata*
27 *Enhydris plumbea*
28 *Fordonia leucobalia*
29 *Chrysopelea ornata*
Elapidae
30 *Naja naja* T
31 *Ophiophagus hannah* E
32 *Bungarus fasciatus* T
Viperidae
33 *Trimeresurus popeorum*
Achrochordidae
34 *Chersydrus granulatus*
35 *Achrochordus javanicus*
Hydrophiidae
36 *Lapenius hardwickii*
Cheloniidae
37 *Chelonia mydas* E
38 *Eretmochelys imbricata* E
39 *Lepidochelys olivacea* V
Crocodylidae
40 *Crocodylus porosus* E

Source: Dat (1997).
a. E, Endangered; V, vulnerable; T, threatened.

Appendix 6.5 Birds in Can Gio Mangrove Biosphere Reserve[a]

Pelecanidae
1 *Pelecanus philippensis* W E R
Ardeidae
2 *Ardea cinerea* W
3 *A. sumatrana* W M
4 *A. purpurea* W
5 *Egretta alba* W
6 *E. intermedia* W
7 *E. garzetta* W
8 *E. sacra* W
9 *Ardeola bacchus* W M
10 *A. speciosa* W
11 *Butorides striatus* W
12 *Bubulcus ibis* W
13 *Nycticorax nycticorax* W
14 *Ixobrychus sinensis* W
Ciconiidae
15 *Mycteria leucocephala* W R
Threskiornithidae
16 *Plegadis falcinellus* W
Anatidae
17 *Dendrocygna javanica* W M
Pandionidae
18 *Pandion haliaetus*
Accipitridae
19 *Elanus caeruleus*
20 *Haliaetus leucogaster*
Rallidae
21 *Rallus striatus* W
22 *Porzana paykullii*
23 *P. fusca*
24 *Amaurornis phoenicurus* W
25 *Gallilula chloropus* W
26 *Porphyrio porphyrio* W
Recurvirostridae
27 *Himantopus himantopus* W
Charadriidae
28 *Vanellus duvaucelii* W
29 *V. indicus* W
30 *Pluvialis squatarola* W M
31 *P. fulva* W
32 *Charadrius dubius* W M
33 *C. alexandrinus* W M
Scolopacidae
34 *Limosa limosa* W M
35 *Numenius phaeopus* W M
36 *N. arquata* W M
37 *Tringa erythropus* W M
38 *T. totanus* W M

39 *T. stagnatilis* W M
40 *T. nebularia* W M
41 *T. ochropus* W M E
42 *T. guttifer* W M R
43 *T. glareola* W M
44 *Actitis hypoleucos* W M
45 *Calidris alba* W M
46 *C. temminckii* W
47 *C. ferruginea* W
48 *Philomachus pugnax* W M
Laridae
49 *Gelochelidon nilotica* W
50 *Hydroprogne caspia* W M
51 *Sterna hirundo* W M
52 *S. dougallii* W
53 *S. sumatrana* W
54 *S. albifrons* W
Columbidae
55 *Streptopelia tranquebarica*
56 *S. chinensis*
57 *Treron fulvicollis*
58 *T. vernans*
Psittacidae
59 *Psittacula alexandri*
Cuculidae
60 *Cuculus micropterus*
61 *Centropus sinensis*
62 *C. bengalensis*
Tytonidae
63 *Tyto alba*
Caprimulgidae
64 *Caprimulgus monticolus*
Apodidae
65 *Aerodramus fuciphagus*
Alcedinidae
66 *Alcedo atthis*
67 *A. meninting*
68 *Halcyon smyrnensis*
69 *H. pileata*
70 *H. chloris*
Meropidae
71 *Merops supercilius*
Pricidae
72 *Picus vittatus*
73 *Chrysocolaptes lucidus*
74 *Dryocopus javensis*
Pittidae
75 *Pitta moluccensis*
76 *P. sordida*

Appendix 6.5 (cont.)

Alaudidae
77 *Alauda gulgula*
Hirundinidae
78 *Hirundo rustica* M
79 *H. daurica*
Motacillidae
80 *Motacilla flava*
Campephagidae
81 *Tephrodornis pondicerianus*
Pycnonotidae
82 *Pycnonotus goiavier*
Irenidae
83 *Aegithina tiphia*
Laniidae
84 *Lanius cristatus*
Turdidae
85 *Copsychus saularis*
86 *C. malabaricus*
87 *Saxicola torquata*
88 *S. jerdoni*
Timaliidae
89 *Pellorneum tickelli*
90 *P. ruficeps*
91 *Garrulax leucolophus*
92 *G. chinensis*
93 *Alcippe peracensis*
Sylviidae
94 *Megalurus palustris*
95 *Acrocephalus bisrigiceps*
96 *A. orientalis*
97 *A. aedon*
98 *Prinia rufescens*
99 *P. hodgsonii*
100 *P. flaviventris*
101 *P. subflava*
102 *Orthotomus sutorius*
103 *O. atrogularis*
104 *O. sepium*

105 *Phylloscopus fuscatus*
106 *P. tenellipes*
107 *P. coronatus*
108 *Cisticola juncidis*
Acanthizidae
109 *Gerygone sulphurea*
Monarchidae
110 *Rhipidura javanica*
Pachycephalidae
111 *Pachycephala grisola*
Paridae
112 *Parus major*
Dicaeidae
113 *Dicaeum concolor*
114 *D. ignipectus*
Nectarinidae
115 *Anthreptes malaccensis*
116 *Nectarinia jugularis*
Zosteropidae
117 *Zosterops palpebrosa*
Ploceidae
118 *Passer montanus*
119 *P. flaveolus*
120 *P. domesticus*
121 *Ploceus philippinus*
Estrildidae
122 *Lonchura punctulata*
Sturnidae
123 *Sturnus nigricollis*
124 *S. burmannicus*
125 *S. sinensis*
126 *Acridotheres grandis*
Dicrunidae
127 *Dicrurus macrocercus*
128 *D. annectans*
Corvidae
129 *Crypsirina temia*
130 *Corvus macrorhynchus*

Sources: Hong et al. (1996); Dat (1997).
a. W, Water bird; M, migratory bird; R, rare; E, endangered.

Appendix 6.6 Mammals in Can Gio Mangrove Biosphere Reserve[a]

Soricidae	Herpestidae
1 *Suncus murinus*	10 *Herpestes urva*
Pteropodidae	11 *H. javanicus*
2 *Macroglossus minimus*	Felidae
Emballonuridae	12 *Felis viverrina* R
3 *Taphazous melanopogon*	13 *Frionailurus bengalensis*
Vespertilionidae	Suidae
4 *Scotophilus heathii*	14 *Sus scrofa*
5 *Miniopterus schreibersi*	Manidae
Cercopithecidae	15 *Manis javanica*
6 *Macaca fascicularis*	Muridae
Mustelidae	16 *Rattus norvegicus*
7 *Lutra lutra* T	17 *R. argentiventer*
8 *Aonyx cinerea* T	18 *Mus musculus*
Viverridae	Hystricidae
9 *Paradoxurus hermaphroditus*	19 *Acanthion subcristatum*

Source: Dat (1997).
a. T, Threatened; R, rare.

REFERENCES

Aksornkoae, S. 1993. *Ecology and Management of Mangroves*. Bangkok: The IUCN Program, 69–70.

Alongi, D.M. and A. Sasekumar. 1992. "Benthic communities." In: A.I. Robertson and D.M. Alongi (eds). *Tropical Mangrove Ecosystems. Coastal and Estuarine Studies*. Washington DC: American Geophysical Union, 137–139.

Anon. 1998. UNESCO-MAB. Regional workshop on Can Gio mangrove biosphere reserve preparation and biosphere reserve networking in the southeast Asian Region, Ho Chi Minh City, 6–11 November 1998. Mangrove Ecosystem Research Division (MERD)/CRES/VNU, 28–35.

Chapman, V.J. 1975. "Mangrove biogeography." In: G.E. Walsh, S.C. Snedaker, and H.J. Teas (eds). *Proc. Int. Symp. Manage. Mangr, 8–11 October 1974, East–West Center, Honolulu*. Vol. 1, 3–22.

Cuong, Nguyen Dinh. 1994. "Management and protection of mangrove forests at Can Gio, Ho Chi Minh City." In: *Proceedings Nat. Work. Refor. Affor. Mang. in Vietnam, Can Gio, Ho Chi Minh City, 6–8 Aug. 1994*, 42–43.

Cuong, V.V. 1964. "Flore et vegetation de la mangrove de la region de Saigon, Cap Saint Jacques." DSc thesis. Paris, 110–120.

Dat, H.D. 1997. "Investigation of mangrove fauna at Can Gio, Ho Chi Minh City." Final report of the project on protection and sustainable use of biodiversity of mangrove ecosystem in Dong Nai Estuary. HCM City: Department of Sciences, Technology and Environment, 45 pp. (in Vietnamese).

Hong, P.N. 1991. "Mangrove vegetation of Vietnam." DSc thesis, Hanoi (in Vietnamese), pp. 35–40.

Hong, P.N. 1996. "Restoration of mangrove ecosystems in Vietnam: a case study of Can Gio District, Ho Chi Minh City." In: Colin Field (ed.). *Restoration of Mangrove Ecosystems*. Okinawa, Japan: The International Tropical Timber Organization and the International Society for Mangrove Ecosystems, 76–79.

Hong, P.N. (ed.). 1997a. *Assessment of the Damages Caused by Chemical War on Mangrove Forests in Vietnam. Final Report of the National Project: "Assessment of the damages caused by chemical war on nature in Vietnam."* Hanoi: National Environment Agency, 55 pp. (in Vietnamese).

Hong, P.N (ed.). 1997b. *The Role of Mangroves in Vietnam – Techniques for Planting and Tending*. Agricultural Publishing House, 206 pp. (in Vietnamese).

Hong, P.N. and H.T. San. 1993. *Mangrove of Vietnam*. Bangkok: IUCN, 46–50, 145–148.

Hong, P.N. and N.H. Tri. 1986. "Effect of herbicides on mangrove forests in the Camau Peninsula. Possibility of forest restoration." In: *"Bakawan"* – Newsletter of the Regional Mangrove Information Network for Asia and the Pacific, December 5(4): 9–10.

Hong, P.N., Nhuong, and L.D. Tuan. 1996.

Hussain, Z. and Gayatri Acharya (eds). 1994. *Mangroves of the Sundarbans*, Vol. 2. Bangladesh: The IUCN Programme, 115–132.

Lai, Bui. 1997. "Shrimp rearing job at Duyen Hai District." Paper presented at the scientific seminar, Ho Chi Minh City, Nov. 1989, 10 pp.

Mien, P.V., B. Lai, D. Canh, D.B. Loc, and P.B. Viet. 1992. "Characteristics of the ecosystem at Sai Gon–Dong Nai estuaries." Paper presented at the scientific seminar, Ho Chi Minh City, Aug. 1992, 7 pp. (in Vietnamese).

Ministry of Sciences, Technology, and Environment. 1992. *The Red Data Book of Viet Nam. Animals*, Vol. 1, 8–24 (in Vietnamese).

Nam, V.N. and T.V. My. 1992. *Mangrove Protection. A Changing Resource System: Case Study in Can Gio District, South Vietnam. Field Doc. No. 3*. Bangkok: FAO, 13–18.

Nam, V.N. and N.D. Quy. 2000. "The need for local participation in policy-making and contract allocations for the mangrove forest protection in Can Gio." In: *Proceedings of the Scientific Workshop: "Management and sustainable use of natural resources and environment in coastal wetlands," Hanoi 1–3 Nov. 1999*, 135–140 (in Vietnamese).

Nhuong, Do Van. 2000. "Benthos in Can Gio Mangrove Forests and trend to sustainable use." Paper presented at the workshop on biodiversity in coastal areas, Hanoi, 15–16 March 2000. Institute for Ecology and Biological Resources: 10 pp. (in Vietnamese).

Ross, P. 1975. "The mangrove of southern Vietnam: the impact of military use of herbicides." In: G.E. Walsh, S.C. Snedaker, and H.J. Teas. (eds). *Proc. Int. Symp. Manage. Mangr.*, 8–11 October 1974, East–West Center, Honolulu. Vol. 2, 3–22, 695–700.

Tang, V.T. 1984. "Initial data of fish fauna in mangrove forests of Minh Hai Province." In: *The First National Symposium on Mangrove Ecosystems, Hanoi, 27–28. Dec. 1984*. Hanoi: Hanoi University of Education, Vol. 2, pp. 310–316 (in Vietnamese).

Tu, L.V. 1996. *Agricultural Land of Ho Chi Minh City and Trend of Utilization.* Ho Chi Minh City: Agricultural Publishing House, 58–61 (in Vietnamese).

Tuan, L.D. 1998. "Sustainable management of the mangrove ecosystem in Can Gio, Ho Chi Minh City." In: P.N. Hong, N.H. Tri, and Q.Q. Dao (eds). *Proceedings of the CRES/MacArthur Foundation of Coastal Biodiversity in Vietnam, Halong City 24–25 Dec. 1997,* 27–39.

Viet, V.Q. 1993. *Report on the Current Situation and the Surface Water at Can Gio District. Final Report of Project No. 69/MT: Environment of Can Gio District.* Ho Chi Minh City: Department of Sciences, Technology and Environment, 7–8 (in Vietnamese).

7

Below-ground carbon sequestration of mangrove forests in the Asia-Pacific region

Kiyoshi Fujimoto

Introduction

Mangrove forests usually create thick, organically rich sediments as their substrata. Most of these substrata in the tropics (except under deltaic environments) consist of mangrove peat which is derived mainly from mangrove roots (Fujimoto and Miyagi 1993; Fujimoto et al. 1995, 1996; Scholl 1964a, b; Woodroffe 1981). This shows that mangrove forests have great below-ground productivity and play a significant role in carbon sequestration, not only above but also below ground.

There have been many studies on the above-ground biomass and productivity of mangrove forests (e.g. Briggs 1977; Christensen 1978; Day et al. 1987; Golley, Odum, and Wilson 1962; Kusmana et al. 1992; Lee 1990; Lugo and Snedaker 1974; Saintilan 1997a, b; Suzuki and Tagawa 1983; Tamai et al. 1986), but relatively few on below-ground biomass (Briggs 1977; Golley, Odum, and Wilson 1962; Komiyama et al. 1987; Lugo and Snedaker 1974; Saintilan 1997a, b). Twilley, Chen, and Hargis (1992) estimated the below-ground carbon storage on the basis of restricted data on the below-ground biomass, excluding the sediments. In this chapter I discuss carbon storage in the sediments of various types of mangroves from different sites of the Asia-Pacific region and the burial rates using the data obtained at Pohnpei Island in Micronesia (Fujimoto et al. 1999a), the south-western coast of the Malay Peninsula in Thailand, Iriomote Island in south-western Japan, and the Mekong River delta in Viet Nam.

Study areas: Methods

Pohnpei Island is situated in the humid tropics without a clear dry season (latitude $6°45'-7°00'$N, longitude $158°05'-158°20'$E). The mangrove forests grow on the coral reef fringing the island and some are situated in estuaries. The former is referred to as coral-reef type and the latter as estuary type. The sediments of the typical coral-reef type consist of a mangrove peat layer 2 m thick (Fujimoto and Miyagi 1993) and the ground surface is usually covered by a forest with dominant *Rhizophora apiculata*, which includes also *Bruguiera gymnorrhiza*, *Sonneratia alba*, and (sometimes) *Xylocarpus granatum* (Fujimoto et al. 1995; Kikuchi et al. 1999).

Field research in Thailand was conducted in the Khlong Thom lowland (latitude around $7°50'$N, longitude around $99°05'$E) and the Satun lowland (latitude around $6°35'$N, longitude around $100°05'$E), which are situated in estuarine environments in the humid tropics with a weak dry season. The mangrove forests in these areas were divided into six communities – namely, the *Sonneratia alba–Avicennia alba*, *Rhizophora apiculata*, *R. apiculata–Bruguiera* spp., *Ceriops tagal–Xylocarpus* spp., *Lumnitzera littorea*, and *Excoecaria agallocha* communities. These six communities are distributed from the middle to the upper part of the tidal zone and correspond to different types of sediments (Mochida et al. 1999).

Iriomote Island is situated in the subtropics (latitude $24°15'-26'$N, longitude $123°40'-56'$E), and the mangrove forests are interesting because they are among the most northerly in the world. Data for this study were collected from the *Rhizophora stylosa*, *R. stylosa–B. gymnorrhiza*, and *B. gymnorrhiza* communities on the small deltas at the mouth of the Nakama, the Urauchi, and the Shiira River, respectively.

Field research in Viet Nam was conducted at Tam Giang III in the Ca Mau Peninsula at the southern tip of the Mekong delta (latitude around $8°50'$N, longitude around $105°15'$E) and the Can Gio Forest Park in eastern Ho Chi Minh City (latitude around $10°25'$N, longitude around $106°52'$E) in the humid tropics with a clear dry season. Data for this study were collected in a *R. apiculata* pure stand converted from a mixed stand with *R. apiculata* that developed on the flood sediments deposited about 100 years ago at Tam Giang III, and a 20-year-old *R. apiculata* plantation on a former *Sonneratia ovata–Avicennia* spp. habitat in Can Gio that was destroyed by herbicides and defoliants sprayed by the United States during the Viet Nam war.

A hand-operated, piston-type, thin-walled sampler 7.5 cm in diameter was used for sampling undisturbed cores. This type of sampler cannot penetrate those sediments with coarse, live roots and undecomposed, large organic debris. Samples (2 cm thick) for the analysis of carbon content were collected at intervals of 10–30 cm. The remains were used to analyse coarse organic-matter content. The coarse organic material (including live roots) was isolated by wet-sieving using 0.5 mm meshed sieves. Total carbon and nitrogen contents were analysed using a CN corer (YANACO MT-600). Samples collected in Viet Nam were analysed by the oxidation–reduction titration method for carbon content in the laboratory of the Ho Chi Minh City Agriculture and Rural Development Service. Carbon-burial rates were calculated using the calendar year, which was calibrated from radiocarbon age using CALIB 1.3.

Results and discussion

Below-ground carbon storage

Table 7.1 shows the stored carbon in the four sites investigated in the Asia-Pacific region. The stored carbon in a coral reef-type mangrove habitat with dominant *R. apiculata* consisting of a mangrove peat layer 2 m thick in Pohnpei Island was estimated at 130 kg C/m^2 (1,300 t C/ha) and that in an estuary-type habitat reached about 200 kg C/m^2 up to 3.5 m in depth (Fujimoto et al. 1999a). In south-western Thailand, the stored carbon up to 1 m in depth in the *R. apiculata*, *R. apiculata–Bruguiera* spp. or *Xylocarpus* spp.–*C. tagal* forests was estimated to be about 50 kg C/m^2, 39 kg C/m^2, and 27 kg C/m^2, respectively (Fujimoto et al. 2000b). In Iriomote Island, it was estimated to be 50 kg C/m^2 for the *R. stylosa* forest at the mouth of the Nakama River, 42 kg C/m^2 for the *R. stylosa–B. gymnorrhiza* forest at the mouth of the Urauchi River, and 22 kg C/m^2 for the *B. gymnorrhiza* forest at the mouth of the Shiira River. In the Mekong delta, it was estimated to be 33 kg C/m^2 for the *R. apiculata* forest on the flood deposit about 100 years old in the Ca Mau Peninsula and 27 kg C/m^2 for the 20-year-old *R. apiculata* plantation on the former *Sonneratia–Avicennia* habitat in Can Gio.

These results suggest that *Rhizophora* forests have a greater carbon-sequestration ability than other types of mangrove forests.

Carbon-burial rate

Table 7.2 shows the carbon-burial rate of mangrove forests at selected sites in the Asia-Pacific region. The annual carbon-burial rates of the mangrove peat layer in the *R. apiculata*-dominant forest in Pohnpei Island were 43 g C/m^2 (= 0.43 t C/ha) between 792BC and AD248 during the falling sea-level phase, 92 g C/m^2 between AD248 and AD671 during

Table 7.1 Below-ground carbon storage in mangrove forests in the Asia-Pacific region

Habitat type	Place	Dominant species[a]	Sediments	No. of plots	Depth (cm)	Stored carbon (kg/m^2)	Reference
Coral reef	Pohnpei Is., Micronesia	Ra	Peat	3	200	130	Fujimoto et al. (1999a)
Small delta or estuary	South-western Thailand	Ra	Peat	2	100	50.16 ± 1.88	Fujimoto et al. (2000b)
		Ra–B or X	Organic clay	4	100	39.00 ± 5.67	Fujimoto et al. (2000b)
		Ct	Inorganic clay	2	100	27.42 ± 2.25	Fujimoto et al. (2000b)
	Iriomote Is., Japan	Rs	Peaty loam	3	100	49.71 ± 6.96	Unpublished
		Rs–Bg	Organic loam	1	100	420	Unpublished
		Bg	Organic loam	2	100	22.00 ± 3.68	Unpublished
Mega-delta	Ca Mau, Southern Viet Nam	Ra on the flood deposit ca. 100 years old	Clay with peat	6	100	33.47 ± 9.17	Fujimoto et al. (2000a)
	Can Gio, Southern Viet Nam	20-year-old Ra plantation on former So–A habitat	Clay	3	100	26.58 ± 3.82	Fujimoto et al. (2000a)

a. Ra, *Rhizophora apiculata*; Rs, *Rhizophora stylosa*; Bg, *Bruguiera gymnorrhiza*; B, *Bruguiera* spp.; X, *Xylocarpus* spp.; Ct, *Ceriops tagal*; So, *Sonneratia ovata*; A, *Avicennia* spp.

Table 7.2 Carbon-burial rates of mangrove forests in the Asia-Pacific region

Place[a]	Dominant species[b]	C-burial period (calendar years)	Stored C (kg/m^2)	C-burial rate (g/m^2/year)	Comments
Pohnpei Island[1]	Ra	792 BC–AD 248	45.1	43	Falling sea-level phase
	Ra	AD 248–671	39.0	92	Rising sea-level phase
	Ra	AD 671–1997	75.6	57	Stable sea-level phase
South-western Thailand[2]	Ra–X	AD 286–1997	79.5	112	Khlong Thom lowland
	Ra–Bc	AD 398–1996	46.1	77	Satun lowland
	Ra	AD 692–1997	43.4	34	Khlong Thom lowland
	Ra	1033 BC–AD 1996	73.7	24	Khlong Thom lowland
	Ct	AD 409–1998	23.8	15	Satun lowland
Iriomote Island[3]	Rs	AD 452–1999	36.1	66	Nakama River
	Rs–Bg	AD 657–1999	41.7	31	Urauchi River
	Bg	AD 1492–1996	19.9	39	Shiira River
Southern Viet Nam[4]	Ra	AD 1899–1998	22.2	224	On the flood deposit in Ca Mau
	Ra	AD 1978–1998	12.7	633	20-year-old plantation in Can Gio

a. Data sources: [1]Fujimoto et al. (1999a), [2]Fujimoto et al. (2000b), [3]K. Fujimoto (unpublished), [4]Fujimoto et al. (2000a).
b. Ra, *Rhizophora apiculata*; Rs, *Rhizophora stylosa*; Bc, *Bruguiera cylindrica*; Bg, *Bruguiera gymnorhiza*; Ct, *Ceriops tagal*; X, *Xylocarpus* spp.

the rising sea-level phase, and 57 g C/m^2 during the last 1,326 years in the stable sea-level phase. These results suggest that *Rhizophora* forests display their greatest below-ground carbon-sequestration ability when accumulating mangrove peat during the rising sea-level phase within the possible peat-accumulation rate.

In the south-western coast of Thailand, relatively high annual carbon-burial rates were found in the *R. apiculata–Bruguiera* spp. or *Xylocarpus* spp. communities, i.e. 77 g C/m^2 during the last 598 years and 112 g C/m^2 during the last 711 years. Sedimentation rates, about 2 mm/year, were also relatively high (Fujimoto et al. 1999b). The high sedimentation and carbon-burial rates appear to have been due to the large input of both inorganic and organic material from upstream. In the *R. apiculata* community, the annual carbon-burial rates were calculated at 24 g C/m^2 during the last 3,033 years and 34 g C/m^2 during the last 1,305 years. The lowest annual value was obtained from the *C. tagal* community, which was 15 g C/m^2 during the last 1,589 years.

The mangrove forests of Iriomote Island usually have a mangrove organic layer 1 m thick formed during the last 1,000 years (Fujimoto and Ohnuki 1995). The annual carbon-burial rates in Iriomote Island were estimated to be 66 g C/m^2 for the *R. stylosa* forest during the last 547 years at the mouth of the Nakama River, 31 g C/m^2 for the *R. stylosa–B. gymnorrhiza* forest during the last 1,342 years at the mouth of the Urauchi River, and 39 g C/m^2 for the *B. gymnorrhiza* forest during the last 504 years at the mouth of the Shiira River.

The annual carbon-burial rate of the *R. apiculata* forest in the Ca Mau Peninsula developed on the flood deposit was estimated to be 224 g C/m^2 during the last 99 years, assuming that the carbon content of the lower part of the flood deposit having scarce mangrove roots was the initial value in the sedimentary deposition. At Can Gio, the carbon-burial rate during the last 20 years after raising the plantation of *R. apiculata* was estimated to be 633 g C/m^2, assuming that the carbon content of the lower layer with few mangrove roots was the initial value at the time of plantation.

The relationship between carbon-burial period and stored carbon for *Rhizophora* forests shows that the *Rhizophora* forest displays its greater below-ground carbon-sequestration ability during the early stage of the forest (fig. 7.1).

Acknowledgements

I am grateful to Prof. Toyohiko Miyagi, Tohoku-Gakuin University, for coordinating the research project in Thailand and Viet Nam and to Dr

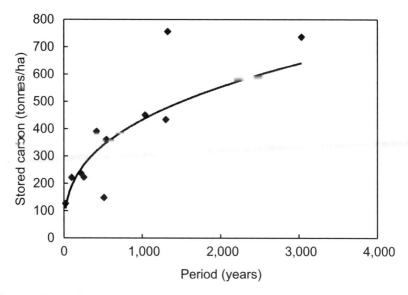

Figure 7.1 Relationship between carbon-burial period and stored carbon for *Rhizophora* forest in the Asia-Pacific region ($y = 37.38x^{0.3544}$; $R^2 = 0.6977$)

Shigeyuki Baba, ISME Executive Secretary, University of the Ryukyus, for enabling me to write this chapter. This study was financially supported by Grant-in Aids for International Scientific Research (Field Research) (FY 1996–1997, No. 08041111 and FY 1998–1999; No. 10041125) and for Scientific Research (B) (FY 1999–2001; No. 11480020) of the Ministry of Education, Science, and Culture of Japan; Special Coordination Funds for Promoting Science and Technology of the Science and Technology Agency of Japan (FY 1996); and Nanzan University Pache Research Subsidy I-A (FY 2000).

REFERENCES

Briggs, S.V. 1977. "Estimates of biomass in a temperate mangrove community." *Australian Journal of Ecology* 2: 369–373.

Christensen, B. 1978. "Biomass and primary production of *Rhizophora apiculata* Bl. in a mangrove in Southern Thailand." *Aquatic Botany* 4: 43–52.

Day, J., W. Conner, F. Ley-Lou, R. Day, and A. Machado. 1987. "The productivity and composition of mangrove forests, Laguna de Terminos, Mexico." *Aquatic Botany* 27: 267–284.

Fujimoto, K. and T. Miyagi. 1993. "Development process of tidal-flat type mangrove habitats and their zonation in the Pacific Ocean: A geomorphological study." *Vegetatio* 106: 137–146.

Fujimoto, K. and Y. Ohnuki. 1995. "Developmental processes of mangrove habitat related to relative sea-level changes at the mouth of the Urauchi River, Iriomote Island, Southwestern Japan." *Quarterly Journal of Geography* 47: 1–12.

Fujimoto, K., R. Tabuchi, T. Mori, and T. Murofushi. 1995. "Site environments and stand structure of the mangrove forests on Pohnpei Island, Micronesia." *Japan Agriculture Research Quarterly* 29: 275–284.

Fujimoto, K., T. Miyagi, T. Kikuchi, and T. Kawana. 1996. "Mangrove habitat formation and response to Holocene sea-level changes on Kosrae Island, Micronesia." *Mangroves and Salt Marshes* 1: 47–57.

Fujimoto, K., A. Imaya, R. Tabuchi, S. Kuramoto, H. Utsugi, and T. Murofushi. 1999a. "Below-ground carbon storage of Micronesian mangrove forests." *Ecological Research* 14: 409–413.

Fujimoto, K., T. Miyagi, T. Murofushi, Y. Mochida, M. Umitsu, H. Adachi, and P. Pramojanee. 1999b. "Mangrove habitat dynamics and Holocene sea-level changes in the southwestern coast of Thailand." *TROPICS* 8: 239–255.

Fujimoto, K., T. Miyagi, H. Adachi, T. Murofushi, M. Hiraide, T. Kumada, M.S. Tuan, D.X. Phuong, V.N. Nam, and P.N. Hong. 2000a. "Belowground carbon sequestration of mangrove forests in Southern Vietnam." In: T. Miyagi (ed.). *Organic Material and Sea-level Change in Mangrove Habitat*. Sendai, Japan: Tohoku-Gakuin University, 30–36.

Fujimoto, K., T. Miyagi, T. Murofushi, H. Adachi, A. Komiyama, Y. Mochida, S. Ishihara, P. Pramojanee, W. Srisawatt, and S. Havanond. 2000b. "Evaluation of the belowground carbon sequestration of estuarine mangrove habitats, Southwestern Thailand." In: T. Miyagi (ed.). *Organic Material and Sea-level Change in Mangrove Habitat*. Sendai, Japan: Tohoku-Gakuin University, 101–109.

Golley, F.B., H.T. Odum, and A.F. Wilson. 1962. "The structure and metabolism of a Puerto Rico mangrove forest in May." *Ecology* 43: 9–19.

Kikuchi, T., Y. Mochida, T. Miyagi, K. Fujimoto, and S. Tsuda. 1999. "Mangrove forests supported peaty habitats on several islands in the Northwestern Pacific." *TROPICS* 8: 197–205.

Komiyama, A., K. Ogino, S. Aksornkoae, and S. Sabhasri. 1987. "Root biomass of a mangrove forest in southern Thailand. 1. Estimation by the trench method and the zonal structure of root biomass." *Journal of Tropical Ecology* 3: 97–108.

Kusmana, C., S. Sabiham, K. Abe, and H. Watanabe. 1992. "An estimation of above ground tree biomass of a mangrove forest in East Sumatra, Indonesia." *TROPICS* 4: 243–257.

Lee, S.Y. 1990. "Primary productivity and particulate organic matter flow in an estuarine mangrove wetland in Hong Kong." *Marine Biology* 106: 453–463.

Lugo, A.E. and S.C. Snedaker. 1974. "The ecology of mangroves." *Annual Review of Ecology and Systematics* 5: 39–65.

Mochida, Y., K. Fujimoto, T. Miyagi, S. Ishihara, T. Murofushi, T. Kikuchi, and P. Pramojanee. 1999. "A phytosociological study of the mangrove vegetation in the Malay Peninsula – special reference to the micro-topography and mangrove deposit." *TROPICS* 8: 207–220.

Saintilan, N. 1997a. "Above- and below-ground biomass of two species of mangrove on the Hawkesbury River estuary, New South Wales." *Marine and Freshwater Research* 48: 147–152.

Saintilan, N. 1997b. "Above- and below-ground biomass of mangroves in a subtropical estuary." *Marine and Freshwater Research* 48: 601–604.

Scholl, D.W. 1964a. "Recent sedimentary record in mangrove swamps and rise in sea level over the southwestern coast of Florida: Part 1." *Marine Geology* 1: 344–366.

Scholl, D.W. 1964b. "Recent sedimentary record in mangrove swamps and rise in sea level over the southwestern coast of Florida: Part 2." *Marine Geology* 2: 343–364.

Suzuki, E. and H. Tagawa. 1983. "Biomass of a mangrove forest and a sedge marsh on Ishigaki Island, South Japan." *Japanese Journal of Ecology* 33: 231–234.

Tamai, S., T. Nakasuga, R. Tabuchi, and K. Ogino. 1986. "Standing biomass of mangrove forests in Southern Thailand." *Journal of the Japanese Forestry Society* 68: 384–388.

Twilley, R.R., R.H. Chen, and T. Hargis. 1992. "Carbon sinks in mangrove and their implications to carbon budget of tropical coastal ecosystems." *Water, Air, and Soil Pollution* 64: 265–288.

Woodroffe, C.D. 1981. "Mangrove swamp stratigraphy and Holocene transgression, Grand Cayman Island, West Indies." *Marine Geology* 41: 271–294.

Part II
Function and management

8

Sustainable use and conservation of mangrove forest resources with emphasis on policy and management practices in Thailand

Sanit Aksornkoae

Introduction

The management of mangrove resources on a sustainable basis is very significant and needs to be implemented seriously. Mismanagement of mangroves will affect negatively not only the mangrove ecosystem proper but also adjoining coastal ecosystems, particularly sea-grass beds and coral reefs as well as the entire coastal system. This is because mangroves are a major component of the tropical coastal belt, with a very important role in the intensive physical, chemical, and biological dynamism of the coastal area. The Government of Thailand recognizes clearly the impacts suffered, and the benefits provided, by the mangroves and has therefore made significant efforts for some time to manage mangrove resources on a sustainable basis; however, several constraints have, hitherto, prevented the full implementation of sustainable management. The critical problems are the conversion of mangrove areas to various purposes for both economic and social development, particularly for shrimp farming and infrastructural development. These activities have caused the loss of large mangrove areas. As a consequence, the extent, quality, and composition of mangrove forests of Thailand have changed very much over the years.

Recently, the Government of Thailand has formulated a distinct policy and defined measures for the management of mangrove resources. Mangrove areas have been classified into two main zones – conservation and development zones. In management practice this means that no activity of any kind and no use of mangrove land is allowed in the conservation zone; only such logging concessions will be granted as can be agreed to operate under strictly controlled technical conditions under law enforcement. There is, at present, encroachment of local people in the mangrove-conservation zone; these people will be removed and corresponding areas will be allocated to them in the development zone. The new living areas will be managed by the Royal Forest Department (RFD) in cooperation with other concerned departments. Zoning for aquaculture in mangrove areas will be clearly identified, based on practical criteria. The protection of mangrove forests will be strictly implemented with adequate manpower and financial support. Research and application of ecological knowledge in managing mangrove resources, with emphasis on income and life-quality improvement for local communities, will be promoted. Good community participation in mangrove conservation and rehabilitation is one of the most important issues, and includes a policy and management plan. The national policy and management practices for sustainable management and conservation of mangrove resources are discussed below, in the hope that these principles may be applicable also to other tropical coastal areas worldwide.

In Thailand, the existing mangrove forest can be found mainly along the coasts of the Andaman Sea and the Gulf of Thailand. Approximately 50 per cent of the 2,614 km of coastline is covered by mangrove forests. The total extent of mangrove forests was estimated to be 167,000 ha in 1996. The best mangrove forest with large trees and high density can be found along the Andaman Sea coast, whereas the mangrove forest around the Gulf of Thailand is only a narrow strip and has largely been converted to shrimp farming, as a resettlement area, and for industrial development.

In Thailand, as in other Asian countries, mangroves are a source of both timber and non-timber forest products which generate livelihood for numerous people. Timber, firewood, poles, wood for charcoal production, crustaceans, fishes, molluscs, and shellfish are the main products of mangrove ecosystems. The economic importance of mangroves for coastal communities is considerable but has not been analysed in detail. Many commercially important marine animals – especially fishes, shrimps, crabs, and various species of molluscs – use mangroves as nursery grounds and shelter during their larval and juvenile stages. The primary nutrient source for these organisms is detritus, which is particulate organic matter derived mostly from decomposing mangrove forest litter. The role of mangrove forests in providing protection against coastal erosion and in maintaining environmental quality in coastal areas is very well recognized in this country.

Unfortunately, the mangrove forests in Thailand have recently been destroyed or degraded at an alarming rate. More than 50 per cent of the total area disappeared during the 35 years from 1961 to 1996 (table 8.1). However, at present, because of recognition of the importance of the mangrove ecosystem, the Government of Thailand has established distinct policies, laws, and regulations and has formulated an action plan in order to achieve conservation and sustainable management of the mangrove resources of the country.

The mangrove ecosystem differs completely from other ecosystems: mangroves are unique and have special highly complex characteristics; they are transitional ecosystems between land and sea and between fresh water and sea water; management of this ecosystem, therefore, has to be treated as a special case. The principal aim of this chapter is to present the national policy and management plan developed for sustainable use based on the various types of mangrove forests in Thailand.

Status and quality of existing mangrove forests in Thailand

Before discussing details of the national policy and management practices for mangrove forests, it is necessary to understand the status and quality of the mangrove forests of the country. The extent, quantity, quality, and composition of mangrove forest has changed greatly in the past 35 years. Following a survey in 1996, Charuppat and Charuppat (1997) estimated the total existing mangrove forest to be approximately 167,582 ha, using Landsat-5^{TM} data. Of the estimated total mangrove-forest area, approximately 80 per cent is found on the west coast, along the Andaman Sea coast, which is peninsular Thailand. However, the mangrove forest in Thailand was approximately 367,900 ha in 1961; thus, 35 years later, the mangrove area has been reduced to little more than half of its original size.

Mangrove-forest areas have been reclaimed for various purposes, such as aquaculture, agriculture, urbanization, salt production, road and transmission-line set-up, mining, construction of ports and harbours, dredging, and the build-up of various industries and power plants. It has been estimated that, between 1987 and 1995, 44,740–74,942 ha of mangrove forests were reclaimed for aquaculture, mainly shrimp-ponds (OEPP 1999). Conflicts over the use of mangrove land became a critical problem. The majority of existing natural mangrove forests at present contain only small mangrove trees, owing to long-term logging concessions. The mangroves originally consisted of a variety of species, including *Rhizophora* spp., *Bruguiera* spp., *Avicennia* spp., *Sonneratia* spp.,

Table 8.1 Changes in the existing mangrove forest in the period 1961–1996

Year	Mangrove area		Change			Average destruction area/year	
	(rai)	(ha)	(rai)	(ha)	(%)	(rai)	(ha)
Before 1961	2,327,800	372,448.00					
1961	2,299,375	367,900.00	28,425	4,548.00	1.23		
1975	1,954,375	312,700.00	354,000	55,200.00	14.81	24,643.00	3,942.88
1979	1,795,675	287,308.00	158,700	25,392.00	6.82	39,675.00	6,348.00
1986	1,227,724	196,435.84	567,951	90,872.16	24.38	81,136.00	12,981.76
1989	1,128,494	180,559.04	99,230	15,876.80	4.27	33,076.00	5,292.16
1991	1,086,381	173,820.96	42,113	6,738.08	1.82	21,056.00	3,368.96
1993	1,054,266	168,682.56	32,115	5,138.40	1.39	16,057.00	2,569.12
1996	1,047,390	167,582.40	6,876	1,100.16	1.04	2,292.00	366.72
Total	–	–	1,280,410	204,865.6	55.76	–	–

Source: Charuppat and Charuppat (1997).

Ceriops tagal, Xylocarpus spp., *Lumnitzera* spp., and *Kandelia candel*. Because some of these species have become rare in some areas, some mangrove forests at present have a low biodiversity.

In order to increase the area under mangrove cover, afforestation and reforestation programmes are under way in new mud-flats, in degraded forests, in abandoned mining areas, and in abandoned shrimp-ponds. Rehabilitation is carried out not only by government agencies but also by the private sector and by local communities, including shrimp farmers and students from universities and schools.

Policy and management practices of mangrove forests

National policy and management practices for mangrove resources are formulated by considering various situations in order to take appropriate action successfully. The important criteria are based on:
- the real situation of existing mangroves;
- the community needs for mangrove resources;
- the existing knowledge of the ecology of mangrove ecosystems;
- the public opinion on the conservation and management of mangrove resources;
- the new constitutional requirements to manage mangrove resources with effective local community participation;
- the utilization of mangrove resources for the people as a whole, rather than for individuals or specific groups;
- the possibility of successful implementation.

Policy and management practices for mangrove land use

Mangrove zonation is one of the first steps to be taken towards controlled and sustainable utilization of mangrove resources. The Thai Cabinet adopted the principle of mangrove zonation, adapted to local conditions, on 15 December 1987, revised in 1998. The entire mangrove area of Thailand has been classified into two zones – the conservation zone and the development zone. Information for the purpose of zonation was obtained from aerial photographs, Landsat images and maps of the Royal Thai Survey Department, National Park maps, and mangrove-concession maps. Data obtained from aerial photographs and field surveys were used to obtain details regarding species composition and topography, structure and density of mangrove forests, as well as patterns of land use for agriculture and industrial activities. The total area of the conservation and development zones, taken together, is shown by province in table 8.2.

Table 8.2 Area (rai)^a of mangrove land-use zones by province, 1998

Province	Conservation zone (rai)	Development zone (rai)	Total
Trad	64,306.25	26,356.25	90,662.50
Chanthaburi	127,775.00	48,406.25	176,181.25
Rayong	15,550.00	12,100.00	27,650.00
Cholburi	481.25	23,425.00	23,906.25
Chacherngsao	493.75	23,887.50	24,381.25
Samut Prakarn	7,237.50	71,618.75	78,856.25
Bangkok	968.75	10,956.25	11,925.00
Nakorn Pathom	25.00	550.00	575.00
Samut Sakorn	7,593.75	168,925.00	176,518.00
Samut Songkram	6,137.50	62,200.00	68,337.50
Petchaburi	5,501.00	67,100.00	72,601.00
Prachuap Kirikhan	1,331.25	6,793.75	8,125.00
Chumpon	42,068.75	24,381.25	66,450.00
Suratthani	18,356.25	55,412.50	73,768.75
Nakorn Si Thammarat	74,312.50	60,793.75	135,106.25
Phatthalung	1,262.50	14,556.25	15,818.75
Songkhla	5,325.50	32,668.75	37,994.25
Pattani	17,412.50	6,525.25	23,937.75
Ranong	165,350.00	3,612.50	168,962.50
Phangnga	236,106.25	11,762.50	274,868.75
Phuket	17,312.50	–	17,312.50
Krabi	237,837.50	11,650.00	246,487.50
Trang	217,962.50	31,368.75	249,331.25
Satun	215,387.50	28,225.00	243,612.50
Total	1,485,795.25	803,275.25	2,289,070.50

a. 1 ha = 6.25 rai.

Measures for land use in the conservation zone

All existing mangrove forests are allocated to the conservation zone. Any change or utilization of the mangrove forest in this zone is absolutely prohibited: the forest must be protected and preserved in its natural state; all forms of utilization must cease.
- Wherever the forest has been damaged, reforestation should be undertaken by the government agencies concerned.
- Newly formed areas within this zone are considered government property, unless proved otherwise, and plantation of mangroves must be undertaken immediately.
- In any place within this zone, where people have permanently settled,

the government agencies concerned must control further encroachments on mangrove land.
- The validity of permits for logging concessions, shrimp farming, tin mining, or other activities must not be extended when the concessions terminate.
- In cases where it is unavoidable for government agencies to implement projects of great economic importance or national security within this zone, they are allowed to do so, provided that they adhere to the Cabinet's resolutions and submit their proposals with Environmental Impact Assessment (EIA) documents to the Cabinet for a final decision.

Measures for land use in the development zones

All conversions of mangrove areas are defined as development zones. The regulations for land utilization in these zones are as follows:
- Government agencies are authorized to rehabilitate any degraded mangrove areas in this zone.
- In any part of the development zone that was formerly illegally developed, the government agencies concerned are obliged to prohibit such activities and to replace mangrove species.
- All laws and regulations should be revised to facilitate reforestation operations in this zone.
- In areas where tin mining is allowed, the agencies concerned have to follow the resolutions and laws formulated by the Cabinet.
- If it is unavoidable for the agencies concerned to use that particular mangrove land, they are allowed to do so following the Cabinet's resolutions.
- Utilization of mangrove land for fisheries, tin mining, cultivation and other activities must be strictly controlled in accordance with proven conservation techniques.
- Before any activity is started in this area, its effects on the ecosystem must be carefully considered.
- Any proposed activity in this area should follow operational procedures with respect to the Cabinet's resolutions and related laws.
- Land use for any purposes in this zone will be rearranged with technical knowledge backup.

Silvicultural-system practices in managing mangrove forests

Natural mangrove forests

The silvicultural system has been revised from time to time in order to suit auction licensing, to promote effective regeneration, and to prevent illegal cutting. The silvicultural system applied is clear-felling in alternate strips. Rotation was set at 30 years with a felling cycle of 15 years. This is practised by dividing the area into 15 coupes (cutting areas), forming an angle of 45 degrees with the coastline, and cutting alternate strips every 15 years, thus giving a rotation of 30 years. Seedlings and saplings of valuable species are left undisturbed in the strips. In situations where sufficient natural regeneration cannot be obtained, enrichment planting is carried out (Aksornkoae 1993). This silvicultural system shows promising results and continues to be used today.

Mangrove plantations

Mangrove plantation in Thailand is carried out either by the RFD or by private individuals. *Rhizophora apiculata* and *R. mucronata* are the two species that are commercially planted on a large scale. *Rhizophora* species are generally planted at a spacing of 1×1 m or 1.5×1.5 m. The cutting rotation time followed on private plantations is fixed at 10 years for firewood and charcoal burning; the government plantations, however, do not yet have any harvesting scheme.

Policy and management practices for mangrove reforestion

Mangrove reforestation aims to conserve natural coastal ecosystems, including the maintenance of most ecological processes, and preserving as much genetic diversity as possible. Timber resources from mangrove reforestation areas can be managed on a long-term sustainable basis. Thailand has defined its programme for large-scale mangrove plantation (Havanond 1994): large areas of degraded mangrove forest, abandoned shrimp-ponds, abandoned mining areas, and new mud-flats are the target areas to be planted. Approximately 1,600–2,000 ha will be planted yearly. So far, mangrove plantations have been operated successfully. Guidelines for planting mangroves in different areas, as mentioned above, are based on the overall growth-performance matrix. The best-performing species for planting, with spacing 1.5×1.5 m in each site, are as follows (JAM 1997):

- New mud-flats: *Rhizophora apiculata* and *R. mucronata*;
- Abandoned shrimp-ponds: *R. apiculata* and *Bruguiera cylindrica*;
- Abandoned tin-mining areas: *Avicennia officinalis* and *Ceriops tagal*;
- Degraded mangrove forests: *Rhizophora mucronata* and *R. apiculata*.

Reforestation is currently operated not only by the government but also by private sectors and local communities, including students from universities and schools. The RFD has established four Mangrove Seed Centres in different parts of the country, to supply mangrove seedlings to the public.

Policy and management practices for aquaculture development in mangrove areas

In Thailand, the conversion of mangrove areas to aquaculture, mainly of the tiger prawn (*Penaeus monodon*), is common along the eastern and south-eastern coasts of the Gulf of Thailand, especially in the provinces of Samut Songkhram, Samut Sakorn, Rayong, Chantaburi, Chumporn, Surat Thani, and Nakhon Sri Thammarat. Apart from the tiger prawn, which is the most popular species and is reared intensively in Thailand, other species such as *Penaeus merguensis* and *P. indicus* are also cultured in a limited area along the coastline of the upper Gulf of Thailand. In order to control the expansion of shrimp farming in mangrove forests and to manage sustainable shrimp culture, the criteria established for shrimp-culture development are as follows:

- Shrimp-ponds should be established only behind the mangrove forests, with an area between the mangrove forest (as a buffer) and the shrimp-pond in the proportion of at least 5:1 (approximately).
- Shrimp-ponds should be zoned according to the local geographical details, ecological significance, and socio-economic aspects.
- Large-scale shrimp farming should be approved on the basis of EIA principles.
- Shrimp farmers should be made responsible for maintaining the quality of the coastal environment at large, and the mangrove resources.
- All shrimp farms should be registered in order to control the operations for culture management effectively.
- An integrated-management system between aquaculture and the conservation of mangrove forests should be widely promoted, and training offered accordingly.
- In the mangrove area, only sites suitable for aquaculture in terms of physical, chemical, geographical, and biological conditions, can be used. In the case of shrimp farming, only an intensive-culture system should be operated.

Policy for mangrove research and applied knowledge

The four main areas of mangrove research are as follows:
- Research leading to self-subsistence and self-reliance on mangrove resources;
- Research leading to increased productivity and value-added benefits from mangrove resources;
- Research leading to good participation of the general public and of the local communities for mangrove conservation;
- Research indicating the potential of mangrove resources.

The RFD has established two mangrove-research centres, one in Trang Province and the other in Ranong Province. The Mangrove Research Centre at Ranong Province has been declared a UNESCO Biosphere Reserve since 1997. This centre provides facilities for scientists from different parts of the world, to implement cooperative research activities. Training programmes on mangrove ecosystems and sustainable-resources utilization are organized for government operators and the public, including local communities and students; equipment for fieldwork is also provided.

Policy and management practices for community participation in mangrove rehabilitation and conservation

- Creation of strong and powerful community organizations:
  - strong local leadership;
  - relationships in the community;
  - role of the community;
  - community organization skills.
- Establishment of network among community organizations:
  - information exchanges;
  - technical assistance.
- Strengthening of relationship between government agencies and community organizations:
  - recognition of importance of community organizations;
  - technical and financial assistance.
- Promotion of integrated conservation and management systems between mangroves and aquaculture:
  - plantations and fish culture;
  - fish-cage culture in mangroves;
  - crab culture in mangroves.
- Training programmes for communities in conservation and rehabilitation of mangrove forests:
  - planting techniques;
  - integrated management systems;
  - co-management resources system;
  - organization management;
  - community forest laws and new constitution.
- Promotion of women's societies in mangrove conservation and rehabilitation programmes.
- Promotion of fundamental research needs on community participation.

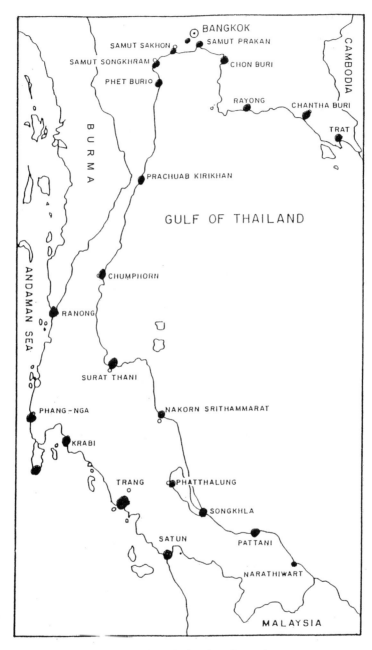

Figure 8.1 Sketch of Thailand, showing the main mangrove areas

Conclusions

In Thailand, mangroves have a very important role, not only for the life and economy of communities of the coastal area but also for the people of the country as a whole. Mangroves provide timber, firewood, poles, wood for charcoal making, thatching material, medicines, fishes, crustaceans, and molluscs. The role of mangroves as spawning ground, nursery, and shelter for various species of commercially important animals and in providing protection against coastal erosion and sea-borne storms is also well known. In the past 35 years, Thailand has lost mangrove areas at an alarming rate: the remaining mangrove area is now only little more than half its original size.

Recognizing the value of mangroves and the dwindling of mangrove areas, Thailand has defined a distinctive national policy and proper practices for the management of mangrove resources. These policies and management practices comprise various important steps and constraints, including mangrove land-use zonation, silvicultural activities, mangrove reforestation, aquaculture development in mangrove areas, mangrove-research programmes, and community participation for mangrove rehabilitation and conservation. It is believed that sustainable use of mangrove resources will be achieved by following the present national policy and rigidly adhering to the management practices advocated by the government.

REFERENCES

Aksornkoae, S. 1993. *Ecology and Management of Mangroves*. Bangkok, Thailand: IUCN, AIT, 176 pp.

Charuppat, T. and J. Charuppat. 1997. *Application of Landsat-5^{TM} of Monitoring the Changes of Mangrove Forest Area in Thailand*. Bangkok, Thailand: Forest Technical Division, Royal Forest Department, 69 pp.

Havanond, S. 1994. "Re-afforestation of mangrove forests in Thailand." *Proceedings of the Workshop on "Development and Dissemination of Re-afforestation Techniques of Mangrove Forests,"* 18–20 April 1994. Bangkok, Thailand: NRCT, 203–216.

JAM. 1997. *Final Report of the ITTO Project on Development and Dissemination of Re-afforestation Techniques of Mangrove Forests*. Tokyo, Japan: Japan Association for Mangroves, 104 pp.

OEPP (Office of Environmental Policy and Planning). 1999. *Estimation of Environmental Costs from Shrimp Farming*. Bangkok, Thailand: Ministry of Science, Technology, and Environment, 6–40.

9

Role of the national government in the economic development of the mangroves of Fiji

Mesake Senibulu

Introduction

The coastal zone where mangroves are found represents an important interface between the land and the sea. For a comparatively small island economy, such as that of the Republic of Fiji, there is a particularly urgent need for utilization of the resources in a sustainable manner. The national government has the primary responsibility in the effective management of mangroves. Provisions in the legislation and government policies are a means of achieving the desired results – that is, the economic development of mangroves.

Fiji is an independent island republic in the South Pacific and lies astride the 180° meridian between longitudes 177°W and 175°E and latitudes 15°S and 22°S. The land area is approximately 18,300 km^2, covering some 320 islands, only about one-third of which are inhabited. The group comprises high islands of volcanic origin and raised coral islands. The two main islands of Viti Levu and Vanua Levu together make up 87 per cent of the total land area.

The island group enjoys a tropical maritime climate with only two seasons – a hot, wet season and a cool, dry season. The hot and wet season runs from November to April and is characterized by variable wind directions and rainfall, as the intertropical convergence zone swings south; tropical cyclones are usually prevalent during this period. The cool and dry season, which occurs from May to October, is determined to a large extent by the steady flow of the prevailing winds, the South-East Trades.

There are some marked variations in rainfall, both from month to month and from year to year. Average figures obtained over a period of 68 years show the capital, Suva, located on a peninsula of the windward side of the main island, to have a mean precipitation of 321 mm/month, whereas Nadi, on the leeward coast, has a mean precipitation of 165 mm/month. The rainfall factor allows for a greater development of rain forest on the eastern side of the main islands.

The population of Fiji at the 1996 census stood at 775,077, comprising indigenous Fijians (51 per cent) and Indo-Fijians (44 per cent), the remaining 5 per cent being made up mainly of Europeans, Chinese, and islanders originating from other Pacific Islands. The principal languages are Fijian and Hindi, although English is the official language.

Distribution and extent of mangroves

The total mangrove area in Fiji has been variably reported as between 19,700 ha (Saenger et al. 1983) and 49,777 ha (Fiji Government). The report by Watling (1985) estimated 38,543 ha for the main islands of Viti Levu and Vanua Levu. However, figures obtained by remote-sensing techniques from the governmental Department of Forestry in their 1999 Annual Report (Conservator of Forests 2000) gave an estimate of 42,464 ha covering the five major islands, including Kadavu, Ovalau, and Gau.

Although mangroves occur also on other islands, their acreage and extent are comparatively small; nevertheless, their sustainable management also needs attention.

In Fiji, detailed scientific information on the fauna and flora of the mangrove ecosystem is scarce, with the result that very little is known about specific physical and chemical processes and their interactions within the coastal waters of Fiji.

Forest types and associated flora

Floristically, Fiji's mangroves are simple, being dominated by only three species and a putative hybrid, all belonging to the family Rhizophoraceae (Richmond and Ackerman 1975). The mangrove species concerned are *Bruguiera gymnorrhiza*, *Rhizophora stylosa*, *R. samoensis*, and *R. selala* (putative hybrid *R. samoensis* × *R. stylosa*).

Four other commonly found tree species and a fern which are also part of the mangrove forests are *Xylocarpus granatum*, *Lumnitzera littorea*,

Table 9.1 Principal species of Fijian mangrove vegetation

Common name	Scientific names
Dogo[a]	*Bruguiera gymnorrhiza* (L.) Lam.
Tiri Tabua[a]	*Rhizophora stylosa* Griff.
Tiri Wai[a]	*R. samoensis* (Hochr.) Salvoza Tomlinson, 1978
Selala[a]	*R.* × *selala* Tomlinson, putative hybrid of *R. samoensis* and *R. stylosa*
Dabi[a]	*Xylocarpus granatum* Koenig
Sagale	*Lumnitzera littorea* (Kack) Voigt
Sinu gaga	*Excoecaria agallocha* L.
Kedra ivi na yalewa kalou	*Heritiera littoralis* Dryand
Borete ferns	*Acrostichum aureum*

a. Dominant species.

Excoecaria agallocha, *Heritiera littoralis*, and *Acrostichum aureum* (table 9.1).

Land ownership and statutory provisions

Land tenure in Fiji divides land into three major classifications, as follows: (1) native land, communally owned by indigenous Fijians and administered on their behalf by the Native Land Trust Board, makes up 83 per cent of the total land; (2) state lands, administered by the Director of Lands on behalf of the state, make up 9 per cent of the total area; (3) the remaining 8 per cent is freehold land.

The principal instruments of statutory legislation relevant to the management of mangroves are the Crown Lands, Fisheries, and Forestry acts.

Land in Fiji has a particularly distinctive characteristic in that, apart from the three major classifications mentioned above, all land below high-water mark belongs to the state, with a special provision of exclusive rights of natives who have traditional rights to fish in these areas.

Role of the national government

Although there is no specific legislation governing the protection of mangroves, the bill that is currently being reviewed by the Parliament of Fiji (the Sustainable Development Bill) will address issues of sustainable management of the country's natural resources, including mangroves.

The national government recognizes the importance of mangroves: in 1933, government designated all mangroves as reserved forests, to be managed by the Department of Forests. However, in 1975 these reserved forests were de-reserved and all mangroves were placed under the jurisdiction of the Department of Lands and Surveys, as an integral part of the foreshore zone (Scott 1993).

The post-independence period saw an increase in foreshore development enterprises: thus, the government (vide Cabinet Decision of 1974) authorized the payment of compensation to traditional fishing-rights owners, because of their loss of fishing rights over the areas reclaimed for development purposes. Although this government action acted to some extent as a restraining factor to would-be foreshore developers, the demand for such locations continued to exert pressure on the processing of proposals.

Concerted efforts by various authorities in Fiji in the early 1980s regarding the sustainable management of mangroves encouraged the government to take affirmative action; this resulted in the establishment of the National Mangrove Management Committee in 1983. This Committee's purpose was to review all foreshore proposals and plans coming as applications to the Department of Lands and to provide recommendations accordingly. A highlight of the work carried out by the Mangrove Management Committee was the production of a National Mangrove Management Plan (Phases I and II) in 1986. This Management Plan was jointly funded by the South Pacific Commission and the Fiji Government. The work is a comprehensive framework intended to assist decision makers in Government, but the drawback is that it still has no legal status.

Department of Lands and Surveys

A function of the Department of Lands and Surveys is the administration of foreshore land under the provision of the Crown Lands Act.

Twelve of thirteen towns and the country's two cities are located in the coastal zone. This situation gives rise to the fact that development pressure would invariably target foreshore land more than any other land type, as it is generally considered easier to procure state land above others. Reclamation for foreshore developments is mainly utilized for agricultural and various urban purposes.

There is a working system for the processing of foreshore proposals. Consultation with other government departments ensures that essential comments and information are available before a decision is taken. This consultation process acts also as a departmental control to regulate foreshore developments.

So far, beginning from the early 1970s, the Department has received more than 1,000 foreshore-reclamation proposals. Those that are processed to completion represent only about 60 per cent of the total; the remainder are either discontinued somewhere in the process or are given development leases but with incomplete developments.

The office dealing with foreshore developments has on hand about 150 applications that have been processed to an advanced stage – that is, between arbitration hearing and issuance of development lease. However, before a lease is granted, a provision in the relevant legislation, dealing especially with the foreshore, requires the public to submit objections, particularly in relation to infringement of public rights. The minister responsible has the final power regarding approval of applications, 75 per cent of which are received from the Central and the Western divisions, which are the location of the major urban centres of the country.

Apart from processing foreshore developments, the Department also deals with harvesting of mangroves, especially *Bruguiera gymnorrhiza* ("dogo") for firewood. Logging licences are issued by the Forestry Department after consultation with the District and Provincial Administration.

Sustainable harvesting

A specific mangrove forest is considered in this report, to illustrate the system used for the economic development of mangroves.

Mangroves of the Rewa delta are found around the mouth of the largest river in Fiji, the Rewa River. The delta is located on the south-eastern part of Viti Levu and east of Suva. According to Scott (1993), the Rewa River delta mangrove forest at present covers an area of 5,130 ha; it is the single largest stand of mangroves, accounting for 12 per cent of the total mangroves of Fiji. Watling (1985) recognizes six main vegetation zones, with about 36 per cent of the total area covered with a uniform closed forest of well-developed *B. gymnorrhiza*. Furthermore, Watling (1985) identifies the Rewa delta mangroves as amongst the most productive in Fiji, and suggests that the primary concern should be to preserve this productivity to sustain subsistence and economic needs.

Primary uses of mangroves, apart from their use as commercial fuel, are as construction materials, fishing devices, dyes, and firewood for domestic use.

The delta, because of its proximity to industrial sites and to the more heavily populated sites on the island, has been the source of some industrial and domestic fuel. Watling (1985) reported that approximately 5,000 m^3 of firewood are extracted annually for domestic use. However, the Forestry Department recently recorded returns of 1,700 m^3 for the 4-year period from 1996, for the use of mangroves mainly as commercial fuel. Firewood suppliers have monthly contracts with a few factories which consider this source as cost-effective for their operations.

A set pattern of harvesting ensures regeneration of seedlings. The system used by the Forestry Department involves leaving aside a mother tree for every 5 m^3 of harvested wood. Over the years, this system has been found to be practicable, and has been made possible through close supervision and monitoring of operations.

Although there is potential for sustainable utilization of mangrove forests for firewood, charcoal, or timber, very little is done to encourage this, mainly because people have not been accustomed to use mangroves for economic gain. It is quite possible that, in the future, when other native timber may become scarce, there may be a demand for mangrove timber. Much research needs to be done in the area of the economic–ecological aspects of mangrove ecosystems in Fiji; there is great potential for research in this area (Lal 1990).

REFERENCES

Conservator of Forests. 2000. "1999 Department of Forestry Annual Report" (unpublished).

Lal, P.N. 1990. *Ecological Economic Analysis of Mangrove Conservation, a Case Study from Fiji. Mangrove Ecosystems Occasional Papers, No 6*. Regional Mangrove Project RAS/86/120. New Delhi: UNDP/UNESCO.

Saenger, P., E.J. Hegerl, and J.D.S. Davie (eds). 1983. *Global Status of Mangrove Ecosystems*. Gland, Switzerland: International Union for Conservation of Nature and Natural Resources (IUCN).

Scott, D.A, (ed.) 1993. *A Directory of Wetlands in Oceania*. Cambridge, UK: IWRB; Kuala Lumpur, Malaysia: AWB.

Watling, D. 1985. *A Mangrove Management Plan for Fiji Phase 1*. Suva: Fiji Government Press.

10

Conflicting interests in the use of mangrove resources in Pakistan

Mohammad Tahir Qureshi

Introduction

The mangroves of Pakistan are a natural resource offering economic, open and ecological, or hidden benefits important for the country. If well managed, mangroves may contribute to the economic development of the coastal area and to the well-being of coastal dwellers. Efforts to utilize this resource have grown, along with increasing demands triggered by human population growth. For the rational utilization of mangrove resources it is necessary to draw detailed plans to avoid conflicts of interests of the various sectors of society involved in the use and management of the coastal zone. It is, therefore, urgent to gain information and experience in all aspects pertaining to the use and management of mangrove ecosystems for future policy-making. Some cases of conflicting interests are briefly mentioned, drawn upon past experience in the field of mangrove management in Pakistan.

Further to the above, the world's seas are under increasing threat from human activities: 3.8 billion people live within 100 km of the coastline; two-thirds of the world's largest cities are coastal; in the next 30 years, more than 6.3 billion people will make their homes in the densely populated coastal areas of the world.

Condition and composition of the mangrove forests

Mangroves in Pakistan cover approximately 260,000 ha, spread over the coasts of the Sindh and Balochistan provinces. Most of the mangroves found in the Indus River delta are well developed and represent the best mangrove forests in the country; mangroves in other parts of the coast either border lagoons or comprise small pockets at the mouth of seasonal rivers. During the past 50 years about 100,000 ha have been destroyed. The rate of destruction of mangrove forests was highest from 1975 to 1992, following the diversion of water of the Indus River to agricultural land upstream, and because of population pressure in general and downstream in particular. At present, natural mangroves exist in limited areas, which will be further reduced as a result of several destructive plans.

Species of mangrove trees present from the marine to the supratidal zone include *Avicennia marina*, *Ceriops tagal*, and *Aegiceras corniculatum*, associated with salt-tolerant non-exclusive mangrove plants. The ecosystem is dominated by a single species, *Avicennia marina*, which represents over 95 per cent of the total crop. *Rhizophora mucronata*, which used to grow in the Indus delta and had been cut down to virtual extinction, has recently been reintroduced successfully (table 10.1; photos 10.1, 10.2).[1] The most abundant species of the fauna found in mangrove forests are invertebrates, various kinds of fish and other marine animals (especially shrimp and crabs), and fish-eating birds such as egrets, cormorants, storks, seagulls, herons, and kites.

Causes and consequences of mangrove conversion

Diversion of fresh water from the Indus River to agricultural land

The aim of agricultural development is to increase food production for self-sufficiency at a national level, for the nutritional improvement of communities, for the increase of employment opportunities, and for the welfare of agricultural workers and fishermen. These targets can be achieved through the application of new forces, such as the intensification or diversification of undertakings. However, the agricultural sector assesses and considers mangroves as wastelands, and the flow of fresh water into mangroves and the sea as wastage (tables 10.2 and 10.3).

The mangrove forests of the delta originated from the mixture of waters of the Indus River with marine tidal waters. The amount of fresh water flowing down the Indus was earlier estimated to be about 150 million acres/foot/year (MAF) or about 180 billion cubic metres (BCM)/year. The delta was formed and grew by the deposition of about 400

Table 10.1 Mangrove species of Pakistan, with their occurrence

	Species	Occurrence
1.	Rhizophoraceae	
	Bruguiera conjugata	Indus delta, Sindh
	Ceriops tagal	Keti Bunder, Miani Hor, Balochistan
	C. roxburghiana	Indus delta, Sindh
	Rhizophora apiculata	Les Bella, Miani Hor, Makran, Balochistan
	R. mucronata	Las Bella, Miani Hor
2.	Myrsinaceae	Indus delta, Sindh
	Aegiceras corniculatum	Miani Hor, Balochistan
3.	Avicenniaceae	
	Avicennia marina	All along the coast of Pakistan
4.	Sonneratiaceae	
	Sonneratia caseolaris	Indus delta, Sindh

Source: Karachi University.

million tons of sediments carried yearly by this amount of fresh water and deposited before the flow reached the sea. Over the last 60 years, humans have been building dams, barrages, and irrigation channels to such an extent that the amount of water reaching the delta is now less than 10 MAF (12 BCM). Proposals and plans being considered indicate that the flow

Table 10.2 Indus River average annual (and seasonal) discharge volumes downstream of Kotri barrage

Period	Discharge volume (BCM)^a			Percentage reduction	Construction (Year)
	Annual	Kharif	Rabi		
1940–1954	101.6	88.5	13.0	10.0	Sukkur barrage (1933)
1955–1965	95.8	83.6	12.2	12.9	Barrages: Kalabagh (Jinnah) (1955) Kotri (1955) Marala (1956) Taunsa (1958) Guddu (1962)
1966–1976	55.2	53.2	1.9	45.7	Warsak dam (1965) Mangla dam (1967) Chashma barrage (1971)
1977–1992	42.2	39.7	2.5	58.4	
1992 onwards (after Water Accord)	12.0				

Source: Irrigation and Power Department, Government of Sindh.
a. BCM, billion cubic metres.

Table 10.3 Percentage salinity in the Indus delta during 1996

Month[a]	Korangi Creek	Keti Bunder	Shah Bunder
January	3.8	3.1	3.1
February	3.8	3.1	3.1
March	3.7	3.0	3.1
April	3.7	2.5	2.5
May	3.5	1.3	1.5
June	3.5	0.4	0.4
July	3.3	0.1	0.1
August	3.5	0.1	0.1
September	3.5	0.1	0.2
October	3.8	2.8	2.9
November	4.1	3.3	3.2
December	4.2	3.8	3.8

a. From May to September, the Indus River inundates the creeks owing to medium to high flood levels.

could be reduced even further because of the additional diversion of fresh water for more land to be brought under agriculture. The quantity of silt carried at present is estimated to be 35 million tons per year.

Reduced river flow is compensated by intrusion of marine waters in the delta tributaries, in channels, in creeks, and in soil pore waters. In future the waters of the Indus River delta may reach even higher salinity levels; even now, it is not unusual to find water salinity of 40–45 parts per thousand (ppt) or more in some parts of the delta, especially in the dry months, already well above normal sea-water salinity. Conditions are such that the growth of trees is negatively affected and animal life is impaired, especially the reproductive and seasonal cycles. In many places, the forest already shows signs of becoming degraded and stunted. The mangrove ecosystem is likely to suffer an even greater impact in the future: there are already signs of stress in seemingly ageing populations of trees, with lack of substantial recruitment in some areas.

Urbanization and industrialization

Destruction of mangroves due to urbanization and industrial expansion is one of the major problems in developing countries. Most decision makers and policy planners have failed to appreciate the utilization and management of mangrove ecosystems as an economically valuable resource. In Pakistan, urbanization and industrial expansion have placed tremendous stress on the mangroves along the Karachi coast. In 1977, Pakistan's second largest port, Bin Qasim, was built to berth ships up to 50,000 tons; it is principally a port for bulk cargo of grains, molasses, oil, iron ore, and other products. Now, there are plans for the expansion of the capacity of Port Qasim to berth ships up to 75,000 tons. The area surrounding the port is being developed into an industrial area, currently dominated by the vast complex of Pakistan Steel Mills, a thermal power plant, and Imperial Chemical Industries (ICI) Pakistan.

Increased accumulation of pollutants in the mangroves, particularly through the mangrove food web, is likely to occur as a result of coastal-development activities. Although the baseline values of various pollutants are at present within the safety concentration limits, we must seriously take into account the fact that aquatic organisms have a strong tendency to accumulate pollutants, especially heavy metals. Moreover, biological modifications in the food web are already evident. Various research institutions have investigated the accumulation of some heavy metals in edible fishes collected from the mangroves. However, according to the National Environmental Quality Standards (NEQS), the heavy metal content of the samples was within the permissible limits and did not pose a threat to the health of humans or of marine organisms.

The levels of marine pollution have been of the following orders of magnitude: 104 million gallons of municipal waste daily; 157 million gallons daily of industrial waste, from tanneries, power plants, steel mills, harbours, ports, etc.; and 1.5 million tons yearly as crude oil pollution.

It is expected that, in the next 10 years, new developments in the tourism sector will open areas adjacent to (and including) mangrove areas. The creeks represent an important source for recreation, water sports, and eco-tourism for the city, which has relatively few such resources nearby. Such developments will change the relationship of the local people with the mangroves and will add additional stresses, if developments are not planned sensibly. Conservation of the area for viewing wildlife – such as migrating waterfowl, dolphins, and jackals – is being anticipated to some extent and the idea of a mangrove biosphere reserve is being pursued.

Overexploitation of mangroves

The mangroves themselves are used by the coastal villagers directly for firewood and fodder for domestic animals. Although *Avicennia* wood does not burn as well as other mangrove species such as *Rhizophora*, it is still cropped extensively by local people for their own use. Although it is rarely sold outside the coastal area, along the north edge of the Indus delta about 100,000 people use about 18,000 tons of mangrove firewood each year (tables 10.4 and 10.5).

On the other hand, *Avicennia* leaves are excellent animal fodder and are collected regularly by the villagers. In addition to the village's cattle,

Table 10.4 Frequent use of mangroves for fodder

	Name of area (Goth)														
	Ibrahim Hydari		Ali Akbar Shah		Chasma Goth		Rehri Miani		Lad Basti		Irkanabad		Itehad Colony		Total	
Frequency of use	N[a]	%	N	%	N	%	N	%	N	%	N	%	N	%	N	%
Regular	36	90	–	–	5	100	35	100	4	100	–	–	–	–	80	95
Occasional	4	10	–	–	–	–	–	–	–	–	–	–	–	–	4	5
Total	40	100	–	–	5	100	35	100	4	100	–	–	–	–	84	100

a. N, number of persons surveyed.

Table 10.5 Monthly consumption of mangroves for fuel

Consumption[a] (kg)	Ibrahim Hydari		Ali Akbar Shah		Chasma Goth		Rehri Miani		Lad Basti		Irkanabad		Itehad Colony		Total	
	N	%	N	%	N	%	N	%	N	%	N	%	N	%	N	%
Don't know	3	1.7	1	5.0	2	7	6	6	2	17	–	–	1	25	15	4
Up to 40	14	7.9	1	5.0	3	11	3	3	–	–	1	6	1	25	23	6
41–80	22	12.4	4	20.0	3	11	6	6	1	8	3	19	1	25	40	11
81–120	34	19.2	7	35.0	9	33	19	19	1	8	9	56	1	25	80	22
121–160	38	21.5	3	15.0	6	22	23	23	–	–	2	13	–	–	72	20
161–200	17	9.6	3	15.0	1	4	13	13	1	8	1	6	–	–	36	10
201–250	14	7.9	–	–	1	4	10	10	2	17	–	–	–	–	27	8
251–300	21	11.9	1	5.0	1	4	9	9	4	33	–	–	–	–	36	10
301+	14	7.9	–	–	1	4	11	11	1	8	–	–	–	–	27	8
Total	177	100	20	100	27	100	100	100	12	100	16	–	4	100	356	100

a. Average consumption of mangroves for fuel per household per month = 17.3 kg; maximum consumption = 800 kg.

sheep, and goats, at certain seasons about 16,000 camels are herded into the mangroves (table 10.6). This activity puts considerable pressure on the existing stands of mangroves nearest to the coastal villages, to such an extent that many mature stands are stunted from overgrazing, browsing, and lopping.

Salt-pans

Large areas of mangrove forest have been converted to salt-pans. Approximately 1,000 ha of salt pans are distributed in the mangrove areas, particularly along the Karachi coast. In some places, the owners of salt farms have planted *Avicennia* along the coast line and on the dykes of the salt-pans for protection from wind and sea surges.

Conflicting interests in the use of mangrove resources

Conflicts among mangrove-resource users have always occurred, as no land-use plans were made in the past. Previously, environmental-impact assessment had not been required for the conversion of mangroves to other types of land use, except for forest products: large areas of mangroves were cleared for construction of ports and harbours, for human settlements, and for industries. Management policy for the utilization of mangrove resources was not clearly understood by operators. These problems have led to severe destruction of the mangrove ecosystem in Pakistan, where approximately 37 per cent of the mangroves have been destroyed or converted to other types of land uses, resulting in several ecological and economical losses.

Management plans for sustainable use of mangrove resources

The conversion of mangrove areas to other uses has yearly been on the increase; for this reason, the remaining mangroves should be reserved for conservation purposes or for utilization on a sustained-yield basis, with minimal conversion or destruction of the area. To achieve these purposes, adequate plans for conservation and rehabilitation should be implemented.

In Pakistan, an afforestation/reforestation programme was started in 1985 and, so far, over 16,000 ha have been brought under mangrove plantation throughout the Indus delta and Balochistan coast. In addition, an area of 3,000 ha has been restocked with *Avicennia marina* by assisted

Table 10.6 Estimated number of camels in 1996

Camel browsing	Port Qasim	East Karachi	Keti Bundar	West Shahbundar	Central Shahbundar	East Shahbundar	Total
Permanent	200	1,105	2,035	1,521	2,423	2,124	9,408
Temporary	–	1,116	1,071	2,900	1,400	1,073	7,560
Total	200	2,221	3,106	4,421	3,823	3,197	16,968

natural regeneration. All this has been done through the assistance of the UNDP/UNESCO Regional Mangrove Projects, the International Union for the Conservation of Nature (IUCN), and the Sindh Forest Department budget. Different indigenous and exotic species with some commercial utilization were used in the experimental plantations and it was found from their growth performance that *Avicennia marina* and *Rhizophora mucronata* from the Pakistan coast were the most successful. About 95 per cent of the plantations in the delta is represented by these two species.

A socio-economic survey was conducted in 1990 to investigate the dependence of the coastal population upon mangroves: it was found that 100,000 people were directly dependent upon mangroves in the delta. The inhabitants did not have sufficient fresh water even for drinking purposes; therefore, the social forestry programme of inland species had only a remote chance of success. Through an experimental programme, it was found that a few thousand container plants supplied to the villagers survived in their kitchen gardens, streets, schools, health centres, and similar locations. Through IUCN, a mass-planting programme of mangrove species was then launched in the coastal villages. All these efforts were directed towards creating an alternative source of firewood and fodder, and to reduce the dependence and pressure of villagers on intertidal mangroves. The container plants/propagules of *Avicennia marina* and *Rhizophora mucronata* were supplied to the local villagers, and technical know-how was provided to allow them to grow these plants in the intertidal flats in the vicinity of their villages. Now, crops of mangroves 2–5 years old are growing in the Rehri coastal village and fishermen are managing them for their firewood and fodder requirements. The establishment of these mangrove plantations is a step towards joint forest management, when the community will harvest the crop grown on government land (table 10.7).

Coastal community development in the Indus delta and Miani Hor, Balochistan: a case study

The fishing communities along the coast and the seasonal fishing villages within the mangroves are dependent upon the natural resources of the area, including fish as their major source of income and the mangroves for firewood and animal fodder, as well as for wood in general. As such, mangroves are essential for the livelihood of the people. The coastal dwellers themselves have the greatest impact upon these natural resources, which are, nevertheless, also affected by general pollution and industrial development (table 10.7).

Table 10.7 Rehabilitation of mangroves[a] in the Indus delta and Balochistan coast

Year	Year (Hac)[b]	
1985	(SFD)	100.00
1986	(SFD)	100.00
1987	(SFD)	253.00
1988	(SFD)	430.00
1989	(SFD)	518.00
1990	(SFD)	500.00
1991	(SFD, IUCN)	1,200.00
1992	(SFD, IUCN)	1,000.00
1993	(SFD, IUCN)	1,110.00
1994	(SFD, IUCN)	2,200.00
1995	(SFD, IUCN, ISME)	2,500.00
1996	(SFD, IUCN, ISME)	2,500.00
1997	(SFD, IUCN, Shirkat Gah, ISME)	2,000.00
1998	(SFD, IUCN, Shirkat Gah, ISME)	2,000.00
1999	(SFD, IUCN, Shirkat Gah, ACTMANG)	500.00
2000	(ACTMANG)	10.00

a. Choice of species: 60 per cent *Avicennia marina*, 40 per cent *Rhizophora mucronata, Ceriops tagal, Aegiceras corniculatum*.
b. SFD, Sindh Forest Department; IUCN, International Union for the Conservation of Nature; ISME, International Society for Mangrove Ecosystems; ACTMANG, Action for Mangrove Reforestation.

The causes of the three main problems of the coastal area of Pakistan are overfishing, overgrazing, and overcutting. These are in turn caused by increasing human population pressure, lack of viable alternatives to firewood, and income. In addition, the growing number of animals applies increasing pressure on the mangroves for fodder. There is no control over grazing and cutting of mangroves; the mangrove area of the Indus delta is too large to be effectively patrolled and guarded, even for the joint forces of the Fisheries and Forest departments, although these departments have jurisdiction over the area. They do have the authority, but do not have the means to enforce the law. At present the mangroves are seen as common property; thus, they suffer the same abuses as common land ("the tragedy of the commons"). If this ecosystem is to survive relatively intact, it is imperative that some form of internal management is organized. This can be done only by developing a strategic alliance with the resource users – woodcutters, fodder collectors, and fishermen.

IUCN responded to the need for involvement of the coastal communities in the management of natural resources by establishing a working relationship with the communities in the villages of Rehri, Ibrahim Hydari, and Somniani, with a total of about 150,000 people belonging to various groups. From the outset, the strategy adopted was to identify the most urgent needs of the people and those with most impact on the environment. Formal and informal education was introduced, aiming at making changes that would reduce the pressure upon natural resources – such as improved and more efficient wood-burning stoves, community forestry, kitchen gardens, and village mangrove stands – as well as promoting various aspects of technical training.

An office was set up at Rehri to sponsor such initiatives and to strengthen relationships with villagers. Such projects included organization of a village clean-up, creation of a first-aid post for fishermen, and development of women's handicrafts for income generation, as well as assistance with child care, health, and education, and increased efficiency in domestic cooking in order to save firewood. Community forestry was promoted in collaboration with the Sindh and Balochistan Forestry departments. In addition, the initiative was taken to develop interaction with four community-based organizations (CBOs) which are working at the grass-roots level in four coastal villages along the Karachi coast. An environmental orientation workshop was organized with the community and CBOs at Ibrahim Hydari and Miani Hor; organization of similar workshops in other villages is planned.

At Rehri and Ibrahim Hydari, Jat and Baloch tribes have raised mangrove plantations, measuring 1–2 acres; IUCN provided the seed and the Forest Department trained the villagers to raise mangrove plantations as "wood-lots." Anjuman Samaji Bahbood, a senior Ibrahim Hydari villager, assisted and guided by the Sindh Forest Department and IUCN, has raised 150 ha *Rhizophora mucronata* since 1998 within the limits of Port Qasim. This plantation is a model of management; similar plantations are being raised in other coastal villages with the participation of the communities and CBOs.

This first step towards joint forest management in the coastal zone of Pakistan appears to be very promising and is gaining momentum. It is hoped that, following the creation of awareness among the coastal people, the natural resources will be efficiently managed and the ecosystem will be protected from degradation.

Note

1. Throughout this volume, photographs have been placed at the end of the chapter to which they refer.

FURTHER READING

Abu, F. and P.J. Meynell. 1992. *Sea Level Rise – Possible Impacts on the Indus Delta*. Korangi Ecosystem Project, Paper No. 2.

Asianics. 1999. "Case study on Tarbela Dam" (circulation draft). World Commission on Dams.

Chauperson, H.G., S.K., and G.M. Khattak. 1965. *Forest Types of Pakistan*. Peshawar: Pakistan Forest Research Institute, 87–95.

Field, C.D. (ed.). 1996. *Restoration of Mangrove Ecosystems*. Okinawa, Japan: International Society for Mangrove Ecosystems.

IUCN. 1987. "Rapid assessment of industrial pollution in the Korangi Phitti Creek."

IUCN. 1988. "Proposal of Management Plan for Korangi Phitti Creek, Karachi – Phase II."

Khan, S.A. 1965. "Mangrove forests, their past and present management in Hyderabad region."

Khan, S.A. 1965. *Working Plan of Coastal Zone Afforestation Division from 1964–65 to 1982–83*. Lahore: Agricultural Department, Government of West Pakistan.

Kogo, M., Miyamoto, C. Suda, and M.T. Qureshi. 1987. *Report on Second Consultant Mission for Experimental Plantation for Rehabilitation of Mangrove Forests in Pakistan*. UNDP/UNESCO Reg. Proj. Res. and Training Programme on Mangroves Ecosystem in Asia and the Pacific (Ras/79/002). Tokyo, Japan: Al-Gurm Res. Center.

Qureshi, M.T. 1985. *Country Report on Mangroves in Pakistan*. Karachi: Sindh Forest Department, Government of Sindh.

Qureshi, M.T. 1986. *Working Plan of Mangrove (Coastal Forest) from 1984/85 to 2002*. Karachi: Sindh Forest Department, Government of Sindh.

Qureshi, M.T. 1990. *Experimental Plantation for Rehabilitation of Mangrove Forests in Pakistan*. Third Report UNDP/UNESCO Reg. Project for Research and Training Programme on Mangrove Ecosystem in Asia and Pacific (Ras/86/120). Karachi: Sindh Forest Department, Government of Sindh.

Photo 10.1 *Rhizophora mucronata*: assisted natural regeneration in the Indus delta

Photo 10.2 *Avicennia marina*: plantation in trench system in Miani Hor

11

Mangroves, an area of conflict between cattle ranchers and fishermen

Patricia Moreno-Casasola

Introduction

People have used mangroves for fuel, construction materials, fishing, substances for textiles and leather tanning, food and beverages, and other purposes. Clear-cutting for timber, firewood, and coal production has caused deforestation, as also has conversion to brackish-water ponds for fish and shrimp culture. Managed mangroves are seldom found in Mexico and urban, agricultural, and cattle lands have taken over the space previously occupied by mangroves, disrupting the delicate ecological balance. Indirect threats include contamination by run-off of pesticides, herbicides, and water enrichment from fertilizers used in the fields; eroded sediments have reduced light penetration into the waters; and sedimentation of silt has changed the depth of the lagoons. As of 1990, only about 1 per cent of the world's mangroves received any form of protection, although this figure may be higher today (Hinrichsen 1998). In many cases, however, the enforcement of rules and regulations for the conservation of mangroves is lax and many areas continue to be exploited at unsustainable rates. In Mexico, mangroves have been protected by law for the last five years; nevertheless, mangrove destruction still occurs.

A coastal-management project to create a productive, sustainable environment

The object of this chapter is to discuss a case study in the central Gulf of Mexico, with special emphasis on mangrove conservation and sustainable use. To promote the restoration, conservation, and sustainable use of natural resources along the Gulf Coast of Mexico, we have been working in La Mancha–El Llano as a case study. The site is located on the central coast of the state of Veracruz, 60 km north of the port of Veracruz, in the Gulf of Mexico. We are working with local stakeholders to establish a process by which residents can make informed environmental and economic decisions. The ultimate goal is to develop a community-based resources-management plan and to work actively towards increasing production while restoring and conserving the ecosystems.

The project

The project is conceived as:
- a mechanism to empower local people for the conservation, management, and administration of the resources of their ecosystems;
- a model of interaction, vinculation, and shared goals between the communities and their neighbours, together with the academic and educational institutions that are working in the area and with the government authorities (county, state, and federation);
- a permanent programme that evaluates the environmental condition of the watershed and discusses solutions; it functions as a forum where conflicts among sectors for the use of resources can be discussed and solved;
- a joint effort to conserve and restore the habitats and the ecosystems and ensure continuity of production and productivity;
- a means for ensuring interaction and communication among the people for a more horizontal society.

The project was planned to be implemented in four stages. Stage I puts emphasis on motivating the highest number of people in the community, in legitimatizing the project, elaborating environmental diagnosis, and defining the hierarchy or series of problems. The four instruments used to develop the management plan will help us to reach our objectives in different ways. These are:
1. the watershed management committee;
2. pilot productive projects;
3. land-use planning exercise and programme;
4. conservation and restoration projects based on community action and environmental education.

During Stage II we have been working on generating spaces to increase communication and interaction and to increase abilities. We have established several programmes to help create shared goals and a deeper knowledge of the environment and of the modern society into which people are being thrown. They include an environmental-legislation education programme, an environmental-education programme, a programme for water management and conservation, the development of collaborative actions within the community, and the formalization of agreements. Stage III emphasizes interactions between local inhabitants for creating stronger cooperation mechanisms and local leadership, always having in view the development of a more horizontal society. Finally, Stage IV consolidates local empowerment and the ability of local communities to discuss problems and solutions and to manage their watershed.

The problem of mangrove conservation within the project

The main problem that gave rise to this part of the project was the legal complaint filed by a group of fishermen because a cattle rancher had felled the mangrove all the way down to the lagoon. By Mexican law, this person had to pay a considerable fine; however, that did not help in the recovery of the mangrove nor in the avoidance of similar actions in the future. An analysis of aerial pictures of the lagoons and mangroves in the area, as well as in many other parts of Mexico, showed that the mangrove fringe that surrounded the lagoons was becoming increasingly thinner. We decided to approach the problem in various ways.

Mangroves are at the interface between two systems – water and land. At present, the systems that form this interface can also be interpreted as a natural ecosystem (the lagoon) and a transformed environment in which ranching activities take place. It is an ecologically fragile ecosystem and a conflict area between two social sectors – fishermen and cattle ranchers.

We must understand the problem in depth, not only from an ecological perspective but also according to its social and political repercussions. Mexican coasts cover 11,592 km of beaches, mangroves, estuaries, and rocky shores, surrounded by coral reefs and tropical forests. Despite this extensive coastline, fisheries are not important: our fishing activities represent only 1 per cent of the world capture, with 1,260,000 tons per year. Coastal cities do not show a higher growth rate than inland cities, except those developed for tourism, such as Cancun. Only 17 cities (with more than 50,000 people) are coastal cities, with 5.6 per cent (5,119,581) of the total population of the country. None has more than 750,000 inhabitants. Cattle ranching takes place over 55.2 per cent of Mexico's lands and is the most important primary economic activity (Inegi-Semarnap 1997).

The state of Veracruz has 745 km of coastal ecosystems: these comprise 18 coastal lagoons, 18,162 ha of mangroves (Loa 1994), 27 counties along the coastline, and 2 coastal cities with 10.5 per cent of the state's population. The main economic activity is agriculture and cattle ranching. Tourism (in the county of Veracruz) and the oil and gas industry (in Coatzacoalcos) are additional economic activities. In none of the counties is fisheries the main economic income activity (Inergi-Semarnap 1997); cattle ranchers are, therefore, politically a more powerful sector than fishermen; furthermore, illiteracy is higher among fishermen.

The working area is occupied by a population that migrated from other parts of the state of Veracruz, or from the country at large during the twentieth century. It is not an indigenous population with a century-old experience of resources management. Cattle raising was introduced by the Spaniards and continued by their descendants. During the 1950s and 1960s, Cebu cattle were introduced, and the occupation of the people changed from free-roaming cattle-rearing practices to pasture management. Demand of land for the increasing population has greatly reduced the size of the fields. Fishermen have been fishing for only the past two generations. In the area, the actual fishermen learned from their fathers, who were the first to work the lagoons. That was a time of enormous productivity, when extracting from the environment was the only practice. At that time, many peasants (who also had a few head of cattle) became fishermen. Now, extraction has to be transformed into management.

The community

Ten rural communities are located in the 50 km^2 area of the project, where there are 6,100 inhabitants. There is also a modern housing facility for the engineers working in the local nuclear plant. Some of the communities were established at the end of the nineteenth century, but most are fairly new (1934–1975). Only two communities have drinking water and sewerage facilities; the remainder rely on wells and latrines. Population-structure studies show 18 per cent of the people to be under 10 years of age and 19 per cent between ages 11 and 20; as a result, almost 40 per cent of the people will be in need of jobs in the next years: 18 per cent are between 21 and 31 years old, 15 per cent between 31 and 41; 12 per cent between 41 and 51, and 18 per cent are over 50 years of age. The average number of family members is 3.7–4.5 per household.

There are numerous stakeholders in the area: they include fishermen, peasants, and cattle ranchers, who have lived on the natural resources for many years. Some are organized and some are not; some live in the area and others reside outside the watershed. Still others are late-comers, who see the potential for development of the area; these have started building

Table 11.1 Stakeholders in the area

	Internal	External
Formal	**Fishing Cooperative La Mancha**	**Institute of Ecology (CICOLMA)**
	Fishing Cooperative Tinajitas–El Viejón	Working group
	Fishing Cooperative Dos Barras	County of Actopan
	Beach restaurant owners	SEDUVER (Asuntos Ecologicos)
	Demonstration farm El Farallón	SEDAP – Fisheries Division
	Elementary schools	ZOFEMAT
	County government authorities	PROFEPA
		SEMARNAP
Non-formal	Women, colonia La Mancha	Students and researchers
	Women, El Viejon	Tourism

Bold type indicates groups or individuals with a direct or indirect relation to mangroves.
CICOLMA, Centro de Investigaciones Costeras La Mancha; SEDUVER, Secretaría de Desarrollo Regional Veracruzana; SEDAP, Social and Economic Dimensions of an Aging Population (Research Programme); ZOFEMAT, Zona Federal Maritimo Terrestre [Federal Maritime Land Zone]; PROFEPA, Procuraduría Federal de Protección al Ambiente [Environmental Protection Agency]; SEMARNAP, Secretaría de Medio Ambiente, Recursos Naturales y Pesca [Secretariat of Environment, Natural Resources and Fisheries].

beach houses as a form of development to earn additional income. Table 11.1 shows the different categories. Bold type indicates those that have some connection, direct or indirect, with mangroves.

We believe that the only way to solve the many issues arising is through a holistic approach to the problems, integral solutions, and concerted action by stakeholders. Since we began working on the project, we have promoted a discussion forum where all stakeholders in the community can discuss their problems and state their point of view. Common concerns are pointed out and compromises are reached. This open forum is known as the Management Plan Committee, where stakeholders (fishermen, cattle ranchers, farmers, organized groups of women, restaurant owners, and educational and research institutions) get together with representatives at the three government levels (county, state, and federal) to clarify issues of common interest, find solutions, and define strategies for their implementation. The Management Plan Committee is a setting where stakeholders "distribute and redistribute access to participation in policy making and implementation, and thereby help to maintain or change political and economic relations" (Bryson and Crosby 1993). The Committee meets every two months and has helped to start and sustain a dialogue between fishermen and cattle ranchers over mangrove and field boundaries and other problems, such as using resources for dredging the lagoons, clearing sand bars, and starting a nursery of native plants for restoration activities.

Conserving and managing mangroves

Interfaces are fragile and conflictive zones. At Veracruz, mangroves – in addition to being an interface between land and sea – are also an interface between a natural and a man-made productive system. The conflict between fishermen and cattle ranchers results in environmental degradation. Mangrove conservation and sustainable use has to be approached under an integrated perspective that takes into account social and economic issues, legal aspects, human attitudes and behaviour, political situations, and ecological and ecosystem functioning. Mangrove management, both for protection and for sustainable use, must consider many aspects. Concrete actions and activities have to be discussed and developed with the stakeholders, to ensure an understanding of the role played by mangroves in coastal areas. Our future – the future of humankind – is closely linked with the preservation of the global systems reponsible for mantaining our life as well as that of other life forms (UNESCO 1999).

Programmes and projects, actions, and activities are organized hereunder according to the following headings: society; legislation; education; local economy; politics; and ecosystems ecology.

Society

We are concerned with the people living close to the mangrove: they are directly affected when mangroves are destroyed; they frequently see resources as their main income, and overlook overexploitation. Children and youngsters in rural coastal areas are becoming "urbanized" and are losing contact with (and, therefore, also knowledge of) their environment and resources. The ones closer to the lagoons and the mangrove are the fishermen, frequently with a lower income and at a lower education level than cattle ranchers and even farmers. We have designed several programmes that will help society obtain control over its territory and resources.

Programmes and projects
- Promoting concerted action among sectors for decision making;
- Establishing conservation and/or restoration projects;
- Changing people's attitudes;
- Promoting attitudes among the community for sharing and participating.

Actions
- Developing activities that promote sharing and participating in the mangroves' recovery, such as organizing schoolchildren and teachers, groups of fishermen, and cattle ranchers for the collection and planting of mangrove seedlings;
- Discussing problems and solutions in the management committee, solving conflicts between sectors, and helping to establish mechanisms for promoting cooperation among fishermen and cattle ranchers.

Activities
- Establishment of the Management Plan Committee;
- Promoting establishment of organized groups of people to work in the productive projects, in the conservation projects, and in task groups;
- Practising conflict-solving within and between groups;
- Establishing links at different government levels;
- Promoting activities that demand sharing responsibility, such as picking up litter along the beaches, and learning how watershed water brings us together;
- Organizing workshops for training, for making diagnoses, for promoting group organization, etc.

Legislation

Environmental legislation helps people and government to protect resources; it gives a framework for sustainable use of ecosystems. There is an amazing lack of knowledge, in rural areas, of the importance of environmental legislation, of the role played by government institutions, and of the rights and duties of the people.

Programmes
- Creating and/or increasing a sense of responsibility in the people for the conservation of species, habitats, resources, and ecosystem functioning;
- Using legal tools to promote conservation and restoration of the environment, with involvement of local people and government participation (such as definition of federal marine areas, land planning, and increasing people's knowledge of the environmental legislation and institutions).

Actions
- Creating local citizen committees for the protection of the environment, establishing close links with government agencies in charge of environmental protection;

- Promoting federal agencies to mark, in the field, the limits of the areas – mangroves and beaches – that are under the administration of the federal government;
- Developing a land-planning project with the local people's participation.

Activities
- Training citizen committees (such as courses on environmental law, permits, species traffic, obligations of government institutions);
- Legalizing permits: fishing cooperatives, restaurants located on federal land areas, explaining federal coastal-area demarcation from the legislative and social perspectives, etc.;
- Workshops with each sector to analyse economic activities and land characteristics, protection policies, need for developing better living conditions;
- Defining and planning restoration and reforestation needs together with the community.

Education

Knowledge and education are two of the main ingredients that help people to take control of their lives and resources. Decisions based on the understanding of ecosystem functioning; watershed interrelationships; sensibility to other sectors' needs; and the practice of talking, discussing, and offering convincing arguments; these are mechanisms suitable for creating a more horizontal society within which we may recover the social network that will allow for a better quality of life.

Programmes
- Creating spaces for participation of local inhabitants and promoting their involvement;
- Increasing knowledge about the importance of environmental aspects (biodiversity, ecosystem functioning, etc.);
- Changing attitudes towards the environment and resources exploitation.

Actions
- Developing activities with schoolchildren that will change their view of nature and change their attitudes and level of participation;
- Developing activities and giving information on local resources and watershed functioning to local inhabitants;
- Giving the general public information on the importance of wetlands for everyday life.

Activities
- Creating a special cartoon character "Rizoforín" associated with the mangrove to attract children and people in general;
- Creating material that can easily be used by schoolteachers and schoolchildren in rural communities;
- Organizing visits to coastal ecosystems;
- Organizing a two-day summer field activity with children;
- Creating a theatrical performance in which children participate actively in writing the play;
- Videos on nature themes;
- Organizing an annual Migratory Beach Bird Festival and creating a Wetlands Month.

Local economy

Resource exploitation is the main source of income in rural areas. Local economies have to rely on the sustainability of ecosystems and on the preservation of global systems that support life on Earth. Conservation must take place in everyday life, not only in protected areas.

Programmes
- Establishing productive projects with local groups will help local inhabitants to increase their income, based on sustainable activities.

Actions
- Establishment of a plant nursery to reproduce native species needed in productive, conservation, and restoration projects – among them, mangrove seedlings;
- Recovering wild populations of the blue crab and establishing a habitat conservation and management plan;
- Establishing rustic enclosures for managing fish and crustaceans;
- Promoting and organizing eco-tourism activities.

Activities
- Training courses and workshops for the different activities that have to be undertaken;
- Workshops for promoting group organization and monitoring of advances and problems within each group;
- Training local people in the monitoring of environmental conditions (physicochemistry, water parameters, habitat, etc.);
- Developing marketing strategies.

Politics

Many of the aspects mentioned above rely on policies. Government agencies are one of the most important, as well as stakeholders in rural areas. Their understanding of the need to have a holistic perspective of environmental problems is the basis of success for sustainable development.

Programmes
- Participation of all stakeholders in decision-making;
- Promoting interaction between different productive groups;
- Promoting interaction and common goals among different government levels.

Actions
- Participation of different government levels in the management committee;
- Promoting government participation in solving local problems.

Activities
- Establishment of the Management Plan Committee;
- Training local government officers in environmental issues;
- Lobbying in local, state, and federal government for solving local problems: opening sand bars to allow water exchanges in the lagoons, eliminating garbage deposits close to the lagoons, others.

Ecosystems ecology

Under the head of ecosystems ecology, emphasis was put on species and habitat protection, having in view the protection and restoration of the functioning of the ecosystem, which is seen as the basis of every primary productive activity. Local people should have a better understanding of ecological functions beyond contamination and garbage disposal; this is what ecology means today to the "man in the street."

Programmes
- Research projects for solving local problems;
- Restoration projects seeking solutions applicable to other areas as well.

Actions
- Wetlands functioning;
- Restoration experiments;
- Autoecology of native species;
- Linking all sectors within the watershed, through ecosystem functioning.

Activities
- Research on seed germination, seedling establishment, and vegetative reproduction of native species;

- Restoration of mangrove and freshwater areas degraded by human activities;
- Reforestation of pastures and watershed restoration;
- Agreements among stakeholders for water management and conservation.

Through these programmes, actions, and activities, we are training people to participate in the management plan of the watershed; we are making them more sensitive to the needs of other sectors and to the environmental problems that may affect all of us. We are, further, training people to take care of their own resources and to make decisions. We are working towards sharing our goals. We want to produce, but we also want to protect and restore; we want people to make decisions and to be responsible for our collective future.

REFERENCES

Bryson, J.M. and B.C. Crosby. 1993. *Policy Planning and the Design and Use of Forums, Arenas, and Courts. Environment and Planning B: Planning and Design 20*, 195–220.

Hinrichsen, S. 1998. "Coastal waters of the world." *Trends, Threats and Strategies*. Washington DC: Island Press.

INEGI-SEMARNAP. 1998. *Estadísticas del Medio Ambiente. México 1997. Informe de la situación general en materia de equilibrio ecológico y protección al ambiente, 1995–1996*. Mexico DF: INEGI-SEMARNAP.

Loa, E. 1994. "Los manglares de México; sinopsis general para su manejo." In: D. Suman (ed.). *El ecosistema de manglar en América Latina y la cuenca del Caribe: su manejo y conservación*. New York: Rosenstiel School of Marine and Atmospheric Science, Univ. of Miami-Tinker Foundation, 144–151.

UNESCO. 1999. *Declaration on Science and the Use of Scientific Knowledge. World Conference on Science for the XXI Century: A New Commitment.* UNESCO.

12

Philippine mangroves: Status, threats, and sustainable development

J.H. Primavera

Introduction

The Philippines is an archipelago of more than 7,100 islands bounded by 17,460 km of coastline in South-East Asia. In 1998, the country harvested 2.8 million metric tonnes (t) of products from aquaculture and from municipal and commercial fisheries in 26.6 million hectares of coastal and oceanic waters (BFAR 1998). The major coastal ecosystems are the mangroves, sea-grass beds, and coral reefs.

Status of Philippine mangroves

Tomlinson (1986) lists 54 species of true mangroves (34 major and 20 minor) worldwide (and 60 species of mangrove associates), of which the Philippines harbours around 40 species belonging to 14 families (table 12.1). Among the mangrove sites with high diversity are the island-province of Bohol with 26 mangrove species (Mapalo 1992), Pagbilao Bay in Quezon Province with 24 species (Bravo 1996), Aurora Province with 23 species (Anon. 1996), Ibajay in Aklan Province with 22 species (Primavera 2001), and Puerto Galera, Mindoro and San Remegio, Cebu with 18 species each (Baconguis et al. n.d.; Buot 1994). The fauna are equally diverse: Pinto (1987) recorded 128 fish species from 54 families over an 18-month period in the Pagbilao mangroves. Shorter studies have

Table 12.1 Major and minor[a] mangrove species in the Philippines[b]

Family	Species
I. Acanthaceae	1. *Acanthus ebracteatus*
	2. *A. ilicifolius*
II. Avicenniaceae	3. *Avicennia alba*
	4. *A. officinalis*
	5. *A. marina*
	6. *A. rumphiana*
III. Bombacaceae	7. *Camptostemon philippinensis*
	8. *C. schultzii*
IV. Combretaceae	9. *Lumnitzera littorea*
	10. *L. racemosa*
	11. *L. rosea*[c]
V. Euphorbiaceae	12. *Excoecaria agallocha*
VI. Lythraceae	13. *Pemphis acidula*
VII. Meliaceae	14. *Xylocarpus granatum*
	15. *X. mekongensis*
VIII. Myrsinaceae	16. *Aegiceras corniculatum*
	17. *A. floridum*
IX. Myrtaceae	18. *Osbornia octodonta*
X. Palmae	19. *Nypa fruticans*
XI. Plumbaginaceae	20. *Aegialitis annulata*
XII. Rhizophoraceae	21. *Bruguiera cylindrica*
	22. *B. exaristata*
	23. *B. hainesii*
	24. *B. gymnorrhiza*
	25. *B. parviflora*
	26. *B. sexangula*
	27. *Ceriops decandra*
	28. *C. tagal*
	29. *Kandelia candel*
	30. *Rhizophora apiculata*
	31. *R. lamarckii*
	32. *R. mucronata*
	33. *R. stylosa*
XIII. Rubiaceae	34. *Scyphiphora hydrophyllacea*
XIV. Sonneratiaceae	35. *Sonneratia alba*
	36. *S. caseolaris*
	37. *S. gulngai*[c]
	38. *S. lanceolata*[c]
	39. *S. ovata*

a. Based on Tomlinson (1986).
b. Sources: Brown and Fischer (1920); Arroyo (1979); Fernando and Pancho (1980); Tomlinson (1986); Anon. (1996); Spalding, Blasco, and Field (1997); Yao (1999).
c. N.C. Duke, University of Queensland, pers. com.

recorded 35 fish species belonging to 18 families in North Bais Bay, Negros Oriental (Dolar, Alcala, and Nuique 1991); 35 species from 21 families in Pagbilao, Quezon (Ronnback 1999); and 29 species from 17 families in Cebu (Blok 1995). Mapalo (1997) observed 56 species of birds belonging to 28 families in 11 sites in Central Visayas and Primavera (1998) recorded 9 species of penaeid shrimp in riverine and island mangroves in Guimaras Island.

Mangrove functions and valuation

Aside from fish and shrimp, other animals collected from mangroves are crabs and lobsters, bivalve and gastropod molluscs, and other invertebrates. They may be harvested within mangrove areas (onsite) or in nearshore waters (offsite). A review of mangrove fisheries reveals ranges in production and value (on a per hectare per year basis), respectively, of 13–756 kg and US$91–5,292 for penaeid shrimps, 13–64 kg and $39–352 for mud crabs, 257–900 kg and $475–713 for fish, and 500–979 kg and $140–274 for molluscs (Ronnback 1999). These fish and invertebrates – together with forestry goods or items such as charcoal, firewood, timber, fishing rods, food, beverages, and fodder – comprise the resource function of mangroves. Mangrove-valuation studies give higher values to fishery than to forestry products (table 12.2).

The services provided by mangroves have as much importance as the goods; these services are also referred to as their regulatory function. They cover coastal protection, erosion control, sediment stabilization, flood regulation, nutrient supply and regeneration, treatment of dissolved and particulate wastes, and wildlife habitats. If services are included in the valuation of mangroves, the estimate may increase to more than $10,000/ha/year (table 12.2). Cost–benefit analysis of mangroves of Fiji and Thailand show that net present value (NPV) is highest if most of the mangrove cover is maintained rather than converted to alternative uses (Lal 1990; Pongthanapanich 1996).

The third mangrove function is information – which can be cultural, artistic, or historical. For example, many coastal Philippine towns and villages are named after mangrove shrubs and trees. Foremost among these is the country's premier city of Manila (or Maynila, as the natives call it), whose name comes from the *nilad*, a mangrove scientifically known as *Scyphiphora hydrophyllacea*, which lined the shores of Manila Bay and the banks of the Pasig River in pre-Hispanic times (Primavera 1995). Many Philippine villages are named after mangroves (e.g. Dungon [*Heritiera littoralis*] and Bakhaw [*Rhizophora* spp.] in Iloilo, and Pagatpatan [*Sonneratia* sp.] in Misamis Oriental) and mangrove associates

Table 12.2 Economic values placed on products and services of mangrove systems

Country	Year	Product or service	Value (US$/ha/year)	Reference
Puerto Rico	1973	Complete mangrove ecosystem	1,550	Hamilton and Snedaker (1984)
Trinidad	1974	Complete mangrove system	600	Hamilton and Snedaker (1984)
		Fishery products	125	
		Forestry products	70	
Fiji	1976	Complete mangrove system	950–1,250	Hamilton and Snedaker (1984)
		Fishery products	640	
Fiji	n.d.	Subsistence and commercial fishery products (combined NPV)[a]	5,468	Lal (1990)
		Subsistence and commercial fuelwood (weighted NPV)[a]	164–217	Lal (1990)
Indonesia	1978	Fishery products	50	Hamilton and Snedaker (1984)
Indonesia	1978	Forestry (charcoal, woodchips)	10–20	Hamilton and Snedaker (1984)
Thailand	n.d.	Charcoal production	4,000	McNeely and Dobias (1991)
Thailand	1979	Fish (inside mangroves) and mangrove-assoc. species (caught outside)	130	Christensen (1982)
Thailand	1982	Fish and shrimp	30–2,000	Hamilton and Snedaker (1984)
		Forestry products	30–400	
Brazil	1981–82	Fish (based on extent of open water)	769	Kapetsky (1987)
Malaysia	1979	Shrimp and fish (including estuaries and lagoons)	2,772	Gedney, Kapetsky, and Kuhnhold (1982)
Malaysia	n.d.	Fishery products	750	Ong (1982)
		Forestry products	225	
Malaysia	n.d.	Managed forest (sustained harvest)	11,561	Salleh and Chan (1986)
India	1985	Complete system (including fishery products, maintenance of fauna, air/water purification)	11,314	Untawale (1986)
Philippines	1995	Mangrove products (excluding services)	58	Gilbert and Janssen (1998)
Mexico	1995	Mangrove goods and services	3,854	Cabrera et al. (1998)

Source: After Primavera (2000).
a. Net present value.

(e.g. Balabago or Balibago [*Hibiscus tiliaceus* or *Thespesia populnea*] all over the islands of Luzon and Panay and Talisay [*Terminalia catappa*] in at least 15 islands). The frequent usage of these common names reflects the usefulness and economic importance or ecological impact of the corresponding mangrove species on the daily lives of village people (Gruezo 1999).

Threats

The decline of mangroves from up to 500,000 ha in 1918 (Brown and Fischer 1918) to only 120,500 ha in 1994 (table 12.3) may be traced to overexploitation by coastal dwellers and to conversion to settlements, agriculture, salt-pans and industry (Baconguis, Cabahug, and Alonzo-Pasicolan 1990; Primavera 1995). However, aquaculture remains the major cause of conversion of mangroves to other uses. About half of the 279,000 ha of mangroves lost from 1951 to 1988 were developed into culture ponds (fig. 12.1): 95 per cent of Philippine brackish-water ponds in 1952–1987 were derived from mangroves (PCAFNRRD 1991). Mangrove-to-pond conversion and its attendant socio-economic changes have been documented in detail for the village of Lincod in Maribojoc, Bohol (Ajiki 1985) and for the municipality of Batan in Aklan (Kelly 1996). Inshore or municipal yields of fish and shrimp have been positively correlated with mangrove areas in the Philippines, Malaysia, and Indonesia (Primavera 1995 and references therein). The loss of mangrove cover is reflected by a parallel decline in municipal fish catches, in contrast to increasing brackish-water pond area and production (fig. 12.2).

Pond construction peaked in the 1950s and 1960s at 4,000–5,000 ha/year with government incentives in the form of loans and again with the "Shrimp Fever" of the 1980s (table 12.3). The Fisheries Decree of 1975 (Presidential Decree [P.D.] 704) mandated a policy of accelerated fish-pond development and Administrative Order (A.O.) 125 extended 10-year fish-pond permits and leases to 25 years (table 12.4). Early guidelines on pond construction in mangrove areas (fig. 12.3) provide evidence of this pro-aquaculture and anti-mangrove bias premised on the belief that mangroves and other wetlands are wastelands.

Aside from the promotion of aquaculture as development strategy, the institutional issues affecting Philippine mangroves are low economic rent of mangroves, overlapping bureaucracy and conflicting policies, poor law enforcement, and lack of political will (see Primavera 2000a for a more comprehensive discussion). The government fee for pond lease of $2/ha/year is ridiculously low in contrast to fish and wood harvests estimated at $538/ha/year and $42–156/ha/year, respectively (Schatz 1991). These low

Table 12.3 Total mangrove and brackish-water culture-pond area in the Philippines[a]

Year	Mangrove area (ha)	Brackish-water ponds		Comments[b]
		Total area (ha)	Increase (ha/year)	
1860	no data (n.d.)	n.d.	762 (1860–1940)	First pond recorded in 1863
1918	450,000	n.d.	n.d.	
1940	n.d.	60,998	1,176 (1941–1950)	
1950	418,382 (1951)	72,753	5,050 (1951–1960)	Fish-pond boom: Fisheries Bureau created; IBRD US$23.6 million for pond development
1960	365,324 (1965)	123,252	4,487 (1961–1970)	
1970	288,000	168,118	811 (1971–1980)	Conservation phase: National Mangrove Committee; 79,000 ha mangroves for preservation and conservation
1980	242,000	176,231	4,668 (1981–1990)	Shrimp fever: Commercial availability of fry and feeds; US$21.8 ADB shrimp project
1990	132,500	222,907	3,052 (1991–1994)	
1994	120,500	232,065		

Source: After Primavera (2000).
a. Data sources: Brown and Fischer (1918); Philippine Census (1921) in Siddall, Atchue, and Murray (1985); BFAR (1970, 1980, 1990, 1994); BFD (1970, 1980); NAMRIA (1988); Auburn University (1993); DENR (1996).
b. BFAR, Bureau of Fisheries and Aquatic Research; BFD, Bureau of Forest Development; NAMRIA, National Mapping Resources Information Agency; DENR, Department of Environment and Natural Resources; IBRD, International Bank for Reconstruction and Development; ADB, Asian Development Bank.

fees underprice the rights to harvest public forests and induce mangrove conversion to ponds, but do not penalize low pond production. There is no shortage of administrative decrees, orders, and promulgations to protect remaining mangrove areas and mitigate widespread deforestation (tables 12.4, 12.5). These include strict criteria for designating permanent forests, fish-ponds, and conditions for pond-lease cancellation. However, effective enforcement is hampered by lack of government manpower and resources, overlapping jurisdiction, and bureaucratic corruption. A case in point is the mangrove green belt or buffer zone, which, although

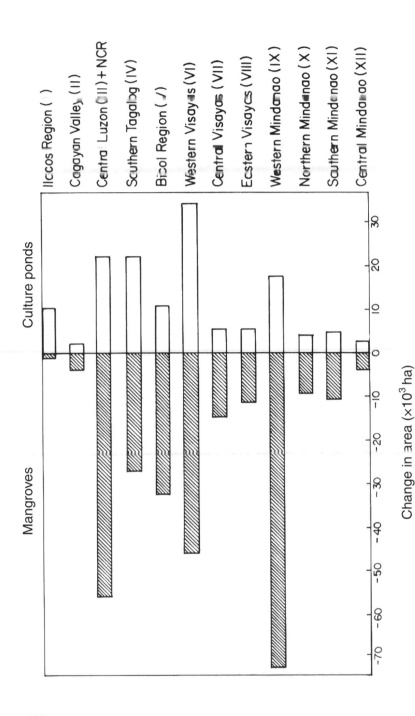

Figure 12.1 Change in area (×10³ ha) of mangroves and brackish-water culture ponds in the Philippines, 1951–1988/1990. Source: Primavera (1997)

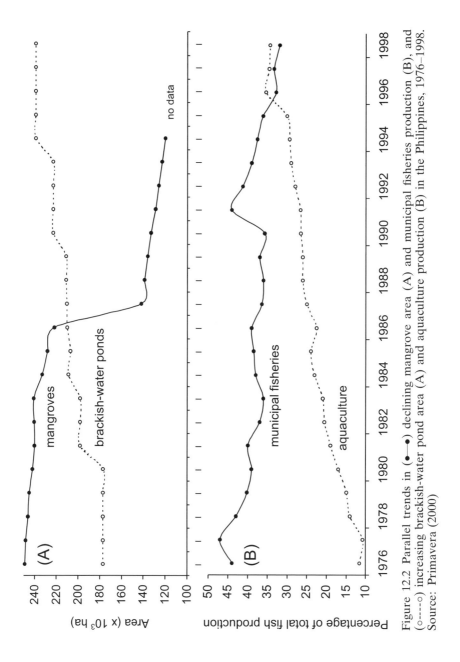

Figure 12.2 Parallel trends in (●——●) declining mangrove area (A) and municipal fisheries production (B), and (o----o) increasing brackish-water pond area (A) and aquaculture production (B) in the Philippines, 1976–1998. Source: Primavera (2000)

Table 12.4 Some Philippine regulations[a] on fish-ponds and mangrove conversion

Regulation	Purpose/description
P.D. 705 (1975)	Fisheries Code: policy of accelerated, integrated fish-pond development; set conditions for mangrove conversion to ponds, public lands for fish-ponds can only be leased, not owned
P.D. 705 (1975)	Revised Forestry Code: retention (and exclusion from pond development) of mangrove strip 20 m wide along shorelines facing oceans, lakes, etc.
P.D. 953 (1976)	Requires fish-pond/mangrove leaseholders to retain or replant 20 m mangrove strip along rivers, creeks
P.D. 1586 (1978)	EIS system (covering resource extractive industries such as fish-ponds)
BFAR A.O. 125 (1979)	Converts fish-pond permits and 10 years FLA to 25 years (to accelerate pond development)
MNR A.O. 3 (1982)	Revises guidelines in classification and zonation of forest lands
DENR A.O. 76 (1987)	Establishes buffer zone: 50 m fronting seas, oceans and 20 m along river-banks; lessees of ponds under FLA required to plant 50 m mangrove strip
R.A. 6657 (1988)	Exempts fish-pond areas from Comprehensive Agrarian Reform Law for 10 years
BFAR A.O. 125-1 (1991)	Increases fish-pond lease from US$2 to US$40/ha/year effective 1992
BFAR A.O. 125-2 (1991)	Delays full implementation of A.O. 125-1
DENR Memo Cir. 7 (1991)	No issuance of mangrove permit for FLA areas with $\geq 10\%$ cover and/or capable of natural regeneration
DENR Gen. Memo Order 3 (1991)	Reverts to forest land all mangrove areas released to BFAR but not utilized or abandoned for 5 years (also under DENR A.O. 15, 1990)
DENR A.O. 34 (1991)	Guidelines for Environmental Clearance Certificate (applicable to fish-ponds)
DENR A.O. 21 (1992)	Implementing guidelines for EIS
R.A. 7881 (1995)	Extends fish-pond exemption from agrarian reform
R.A. 8550 (1998)	Revised Fisheries Code: Pond lessees required to reforest river-banks, bays, streams, and seashore fronting pond dykes

Source: After Primavera (2000).

a. A.O., Administrative Order; BFAR, Bureau of Fisheries and Aquatic Research; DENR, Department of Environment and Natural Resources; EIS, Environmental Impact Statement; FLA, Fish-pond Lease Agreement; MNR, Ministry of Natural Resources; P.D., Presidential Decree; R.A., Republic Act.

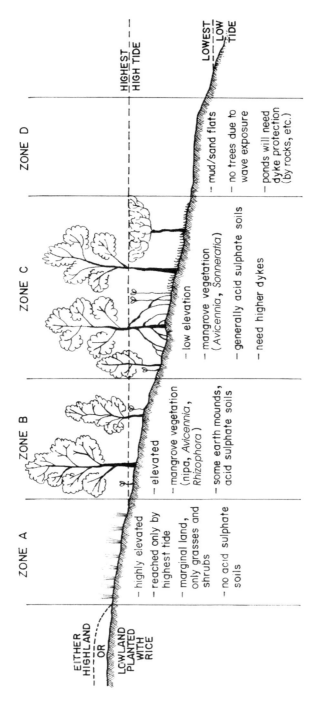

Figure 12.3 Guidelines to mangrove zones as sites for (extensive) aquaculture ponds: note dyke requirements, wave exposure, and acid sulphate soils

Table 12.5 Some Philippine regulations[a] on mangrove conservation and rehabilitation

Regulation	Purpose/description
P.D. 705 (1975)	Revised Forestry Code: Mangrove strips in islands which provide protection from high winds, typhoons shall not be alienated
P.P. 2151 & 2152 (1981)	Declares 4,326 ha of mangroves wilderness areas and 74,767 ha forest reserves
P.P. 2146 (1982)	Prohibits mangrove cutting
BFD Cir. 10 (1986)	Prohibits processing of mangrove lands which are part of forest lands
MNR A.O. 42 (1986)	Expands mangrove forest belt in storm surge, typhoon-prone areas: 100 m along shorelines fronting seas, oceans, 50 m along river-banks
P.D. 1067	3 to 20 m of river-banks and seashore for public use: recreation, floatage, navigation, fishing, and salvage; building of structures not allowed
E.O. 192 (1987)	Mangroves and swamp lands inside public forests under DENR jurisdiction and control
DENR A.O. 77 (1988)	Implementing guidelines of Integrated Social Forestry Program (provides incentives in co-management of forest resources through provision of legal tenure)
DENR A.O. 15 (1990)	Policies on communal forests, plantations, tenure through Mangrove Stewardship Contracts; revert abandoned ponds to forest; ban cutting of trees in FLA areas; prohibit further conversion of thickly vegetated areas
DENR A.O. 123 (1990)	Community Forestry Program (covers relatively large mangrove areas)
DENR A.O. 9 (1991)	Policies and guidelines for Mangrove Stewardship Agreement
R.A. 7160 (1991)	Local Government Code: devolved management/implementation of community forestry projects, communal forests < 500 ha, enforcement of community-based laws
DENR A.O. 13 (1992)	Buffer zone of 50–100 m in storm-prone areas.
DENR A.O. 23 (1993)	Forest Land Management Program: provides 25-year Forest Land Management Agreement of 1–10 ha and 10–1,000 ha to families and communities, respectively
DENR A.O. 30 (1994)	Community-Based Mangrove Forest Management (NGO assistance)

Source: After Primavera (2000).

a. A.O., Administrative Order; BFD, Bureau of Forest Development; DENR, Department of Environment and Natural Resources; E.O., Executive Order; FLA, Fish-pond Lease Agreement; MNR, Ministry of Natural Resources; NGO, Non-governmental Organization; P.D., Presidential Decree; P.P., Presidential Proclamation; R.A., Republic Act.

required by law, is absent in most culture ponds throughout the country. Because of overlapping but ineffective bureaucracy, the forestry and fisheries agencies tasked to administer mangroves and ponds, which share the same resource base, seldom coordinate with each other.

Conclusions and recommendations

The preceding sections illustrate the importance of mangrove ecosystems and their products and services. The question, therefore, is not whether mangroves should be preserved, but how much to preserve. Following the Precautionary Principle, Saenger, Hegerl, and Davie (1983) suggest that no more than 20 per cent of a given mangrove area should be converted into ponds (and/or other uses). However, the state of Philippine mangroves is the reverse of this ideal of one hectare of ponds to four hectares of mangroves. At present, just over 100,000 ha of mangroves are left out of 500,000 ha of mangroves at the turn of the century.

Clearly, these remaining mangroves should be conserved and managed in a sustainable way. The following zones are recommended to be established (Primavera 1993):

- *Protected Forest*: "No Touch" reserve areas of pristine forests for biodiversity maintenance, scientific research and education, fisheries support, and coastal protection;
- *Productive Forest*: sustained yield of forestry and fisheries products;
- *Reforestation Zone*: degraded areas of ecological importance, including abandoned culture ponds, that need to be rehabilitated;
- *Conversion Zone*: to aquaculture, agriculture, salt-pans, etc., not exceeding 20 per cent of total mangrove area, preferably in marginal and previously altered sites.

Mangrove rehabilitation should follow geographical criteria, e.g. planting along seaward coasts of the natural colonizers *Avicennia* and *Sonneratia*, rather than the mortality-prone *Rhizophora*, popularly used for firewood and fishing rods and easily available for reforestation because of large propagules available all year round and easy to plant by pushing the large propagule into the mud. Priority should be given to islands most vulnerable to the 20–30 typhoons that visit the archipelago yearly. In this connection, the green belt or buffer zone 20–50 m wide, required to be planted along river banks, and 50–100 m facing the open sea, as well as other laws protecting mangroves (tables 12.4, 12.5), should be strictly enforced and violators taken to court. Aquaculture operators can be encouraged to use their ponds more efficiently if fees are increased to capture economic rent and provide funds for mangrove rehabilitation (Primavera 2000a).

Mangroves and aquaculture are not necessarily incompatible. Mangrove-Friendly Aquaculture (MFA) can be categorized under two heads: (1) mangrove filters, where mangrove forests are used to absorb effluents from high-density culture ponds, and (2) aqua-silviculture or the low-density culture of crabs, shrimp, and fish integrated with mangroves. The SEAFDEC Aquaculture Department based in Iloilo, Central Philippines, is currently conducting studies in newly regenerating and old-growth mangrove areas using both models (Primavera 2000b).

MFA and mangrove-management projects should be promoted in the context of a wider integrated coastal-zone management (ICZM) that coordinates the needs of various sectors (fisheries, aquaculture, forestry, industry, etc.) and should be community based, in cognizance of the role of local residents as users and day-to-day managers of coastal resources (Primavera and Agbayani 1997).

REFERENCES

Ajiki, K. 1985. "The decrease of mangrove forests and its effect on local peoples' lives in the Philippines." In: Kikuchi, T. (ed.). *Rapid Sea Level Rise and Mangrove Habitat*. Gifa: Institute for Basin Ecosystem Studies, Gifa University, 51–54.

Anon. 1996. *Aurora Integrated Area Development Project II Philippines – A Management and Protection Strategy for Aurora Province*. Baler, Aurora, Philippines: AIADP II Project Management Office, 98 pp.

Arroyo, C.A. 1979. "Flora of the Philippines mangrove." *Biotropica Special Publication* 10: 33–44.

Auburn University. 1993. *Philippine Prawn Industry Policy Study*. Prepared for the Coordinating Council of the Philippines Assistance Program and U.S. Agency for International Development. International Center for Aquaculture Environments, Auburn University.

Baconguis, S.R., D.M. Cabahug, Jr, and S.N. Alonzo-Pasicolan. 1990. "Identification and inventory of Philippine forested-wetlands resource." *Forest Ecology Management* 33/34: 21–44.

Baconguis, S.R., F.T. Ociones, I.A. Panot, R.M. Lavega, F.E. Siapno, C.R. Carino, D.Y. Holgado, and F.M. Reyes. No date. *A Guidebook on the Mangroves of Puerto Galera*. College, Laguna, Philippines: Ecosystems Research and Development Bureau, Department of Environment and Natural Resources.

BFAR. 1970. *Fisheries Statistics of the Philippines*. Quezon City: Bureau of Fisheries and Aquatic Resources, Department of Agriculture.

BFAR. 1980. *Fisheries Statistics of the Philippines*. Quezon City: Bureau of Fisheries and Aquatic Resources, Department of Agriculture.

BFAR. 1990. *Philippine Fisheries Profile*. Quezon City: Bureau of Fisheries and Aquatic Resources, Department of Agriculture.

BFAR. 1994. *Philippine Fisheries Profile*. Quezon City: Bureau of Fisheries and Aquatic Resources, Department of Agriculture.

BFAR. 1998. *Philippine Fisheries Profile*. Quezon City: Bureau of Fisheries and Aquatic Resources, Department of Agriculture.

BFD. 1970. *Forestry Statistics of the Philippines*. Quezon City: Bureau of Forest Development, Department of Environment and Natural Resources.

BFD. 1980. *Forestry Statistics of the Philippines*. Quezon City: Bureau of Forest Development, Department of Environment and Natural Resources.

Blok, L.G. 1995. "Fish community structure and ichthyomass levels in a Philippine mangrove system: an ecological and socio-economic analysis between reforested and natural mangrove in the Central Visayas Region of the Philippines." Unpub. MA thesis, University of Leiden.

Bravo, D.R. ca. 1996. "Restoration and management of the Pagbilao Mangrove Genetic Resource Area." In: C. Khemnark (ed.). *Ecology and Management of Mangrove Restoration and Regeneration in East and Southeast Asia*. Proceedings of Ecotone IV, 18–22 January 1995, Surat Thani, Thailand. Jakarta: National MAB Committee of Thailand, National MAB Committee of Japan, and UNESCO Regional Office for Science and Technology.

Brown, W.H. and A.F. Fischer. 1918. *Philippine Mangrove Swamps. Bureau of Forestry Bulletin No. 17*. Manila: Department of Agriculture and Natural Resources, Bureau of Printing, 132 pp.

Brown, W.H. and A.F. Fischer. 1920. "Philippine mangrove swamps." In: W.H. Brown (ed.). *Minor Products of Philippine Forests I. Bureau of Forestry Bull. No. 22*. Manila: Bureau of Printing, 9–125.

Buot, I.E., Jr. 1994. "The true mangroves along San Remegio Bay, Cebu, Philippines." *Philippine Scientist* 31: 105–120.

Cabrera, M.A., J.C. Seijo, J. Euan, and E. Perez. 1998. "Economic values of ecological services from a mangrove ecosystem." *Intercoast Network* No. 32, Fall 1998.

Christensen, B. 1982. Management and utilization of mangroves in Asia and the Pacific. *FAO Environmental Papers* 3: 1–60.

DENR. 1990. *Compilation of Mangrove Regulations*. Quezon City: Coastal Resources Management Committee, Department of Environment and Natural Resources, 47 pp.

DENR. 1996. *The Philippine Environmental Quality Report, 1990–1995*. Quezon City: Environmental Management Bureau, Department of Environment and Natural Resources.

Dolar, M.L., A.C. Alcala, and J. Nuique. 1991. "A survey of the fish and crustaceans of the mangroves of the North Bais Bay, Philippines." In: *Proceedings of the Regional Symposium on Living Resources in Coastal Areas*, 30 Jan.–1 Feb. 1989, Manila, Philippines. Quezon City: Marine Science Institute, University of the Philippines, 513–519.

Fernando, E.S. and J.V. Pancho. 1980. "Mangrove trees of the Philippines. Sylvatrop Philipp." *Forestry Research Journal* 5(1): 35–54.

Gedney, R.II., J.M Kapetsky, and W.W. Kuhnhold. 1982. *Training on Assessment of Coastal Aquaculture Potential, Malaysia*. Manila: South China Sea Fisheries Development and Coordinating Programme, SCS/GEN/82/35, 62 pp.

Gilbert, A.J. and R. Janssen. 1998. "Use of environmental functions to commu-

nicate the values of a mangrove ecosystem under different management regimes." *Ecological Economics* 25: 323–346.
Gruezo, W.S. 1999. "Of Philippine plants and places: an ethnobotanical memoir." *Asia Life Sciences* 8(1): 15–47.
Hamilton, L.S. and S.C. Snedaker (eds). 1984. *Handbook for Mangrove Area Management*. Honolulu, Hawaii: United Nations Environment Programme, Kenya and Environment and Policy Institute, East-West Center, 123 pp.
Kapetsky, J.M. 1987. Conversion of mangroves for pond aquaculture: some short-term and long-term remedies. *FAO Fish. Rep. Suppl.* 370: 129–141.
Kelly, P.F. 1996. "Blue revolution of the red herring? Fish farming and development discourse in the Philippines." *Asia Pacific Viewpoint* 37: 39–57.
Lal, P.N. 1990. *Ecological Economic Analysis of Mangrove Conservation: A Case Study from Fiji. Mangrove Ecosystems Occasional Papers No. 6.* New Delhi: UNDP/UNESCO Regional Mangroves Project RAS/86/120.
Mapalo, A.M. 1992. "Mangrove species distribution in Bohol, Philippines." *Ecosystems Research Digest* 3(2): 55–62.
Mapalo, A.M. 1997. "Survey of mangrove associated wildlife in Central Visayas." *Ecosystems Research Digest* 7(2): 1–23.
McNeely and Dobias. 1991. "Economic incentives for conserving biological diversity in Thailand." *Ambio* 20(2): 86–90.
NAMRIA. 1988. *Land Condition Mapping Project of the Philippines, 1988 (Manual Interpretation of Spot Satellite Data)*. Makati, Metro Manila: National Mapping Resources Information Agency.
Ong, J.E. 1982. "Mangroves and aquaculture in Malaysia." *Ambio* 11: 252–257.
PCAFNRRD. 1991. *The Philippines Recommends for Mangrove Production and Harvesting, Forestry Research Series No. 74*. Los Baños, Laguna: Philippine Council for Agriculture, Forestry and Natural Resources Research and Development, Department of Science and Technology, 96 pp.
Pinto, L. 1987. "Environmental factors influencing the occurrence of juvenile fish in the mangroves of Pagbilao, Philippines." *Hydrobiologia* 150: 283–301.
Pongthanapanich, T. 1996. "Applying linear programming: economic study suggests management guidelines for mangroves to derive optimal economic and social benefits." *Aquaculture Asia* 1(2): 16–17.
Primavera, J.H. 1993. "A critical review of shrimp pond culture in the Philippines." *Review of Fishery Science* 1: 151–201.
Primavera, J.H. 1995. "Mangroves as brackishwater pond culture in the Philippines." *Hydrobiologia* 295: 303–309.
Primavera, J.H. 1997. "Socioeconomic impacts of shrimp culture." *Aquacultural Research* 28: 815–827.
Primavera, J.H. 1998a. "Mangroves as nurseries; shrimp populations in mangrove and non-mangrove habitats." *Est. Coastal Shelf Science* 46: 457–464.
Primavera, J.H. 1998b. "Tropical shrimp farming and its sustainability." In: S. De Silva (ed.). *Tropical Mariculture*. London: Academic Press, 257–289.
Primavera, J.H. 2000a. "Development and conservation of Philippine mangroves: institutional issues." *Ecological Economics* 35: 91–106.
Primavera, J.H. 2000b. Integrated mangrove–aquaculture systems in Asia. *Integrated Coastal Zone Management* (Autumn) 121–130.

Primavera, J.H. 2001. "The patch mangroves of Ibajay, Aklan in Panay Island – a microcosm of Philippine mangroves." UNESCO–MAB Ecotone X on Ecosystem Valuation. Hanoi, Viet Nam, 19–23 November 2001.

Primavera, J.H. and R.F. Agbayani. 1997. "Comparative strategies in community-based mangrove rehabilitation programmes in the Philippines." In: P.N. Hong, N. Ishwaran, H.T. San, N.H. Tri, and M.S. Tuan (eds). *Proceedings of Ecotone V, Community Participation in Conservation, Sustainable Use and Rehabilitation of Mangroves in Southeast Asia*. Viet Nam: UNESCO, Japanese Man and the Biosphere National Committee and Mangrove Ecosystem Research Centre, 229–243.

Ronnback, P. 1999. "The ecological basis for economic value of seafood production supported by mangrove ecosystems." *Ecological Economics* 29: 235–252.

Saenger, P., E.J. Hegerl, and J.D.S. Davie (eds). 1983. *Global Status of Mangrove Ecosystems. IUCN Commission on Ecology Papers No. 3*. Gland, Switzerland: IUCN.

Salleh, M.N. and H.T. Chan. 1986. "Sustained yield forest management of the Matang mangrove." In: *Mangroves of Asia and the Pacific: Status and Management. Tech. Rep. UNDP/UNESCO Research and Training Pilot Programme on Mangrove Ecosystems in Asia and the Pacific (RAS/79/002)*. Philippines: Natural Resources Management Center and National Mangrove Committee, Ministry of Natural Resources, 319–324.

Schatz, R.E. 1991. *Economic Rent Study for the Philippine Fisheries Sector Program*. Manila: Asian Development Bank TA 1208-PH1, 42 pp.

Siddall, S.E., J.A. Atchue, III, and P.L. Murray, Jr. 1985. "Mariculture development in mangroves: a case study of the Philippines, Panama and Ecuador." In: J.R. Clark (ed.). *Coastal Resources Management: Development Case Studies, Renewable Resources Information Series, Coastal Management Pub. No. 3*. Prepared for the National Park Service, U.S. Dept. of the Interior, and the U.S. Agency for International Development. Columbia, South Carolina: Research Planning Institute, Inc.

Spalding, M., F. Blasco, and C. Field (eds). 1997. *World Mangrove Atlas*. Okinawa: International Society for Mangrove Ecosystems, 178 pp.

Tomlinson, P.B. (ed.). 1986. *The Botany of Mangroves*. Cambridge: Cambridge University Press, 413 pp.

Untawale, A.G. 1986. "Country reports: India." In: *Mangroves of Asia and the Pacific: Status and Management, Tech. Rep. UNDP/UNESCO Research and Training Pilot Programme on Mangrove Ecosystems in Asia and the Pacific (RAS/79/002)*. Philippines: Natural Resources Management Center and National Mangrove Committee, Ministry of Natural Resources, 51–87.

Yao, C.E. 1999. "Bakawan hybrid, the fourth *Rhizophora* species in the Philippines?" *Tambuli* May, 19–20.

13

Co-management of coastal fisheries resources in tropical and subtropical regions

Shinichiro Kakuma

Introduction

Coastal (coral or mangrove area) fisheries resources in the Pacific Islands, South-East Asia, and Okinawa recently have been overexploited and depleted. In the Pacific Islands in particular, one of the causes of this depletion has been the recent population increase of 2.5–3.5 per cent (United Nations 1997). A great demand for fish gave impetus to the overfishing: the demand greatly exceeded local production and the islanders had to import canned fish (an amount more than half the local production). The introduction of underwater torches and SCUBA dramatically increased fishing efficiency, and the destruction of nursery areas by human activities resulted in decrease of the fisheries resources (King and Fa'asili 1997). The use of explosives and chemicals such as cyanide for live reef fish severely affected the inshore ecosystem.

To tackle these problems, a variety of management methods have been tried. Most of these tended to be government-based, centralized, top-down approaches. Such approaches (which evolved in Western developed countries), however, did not work effectively in the tropics; conversely, community-based management or co-management is considered to be more suited to the regions (Adams 1996; Flores and Marte 1996; Johannes, Ruddle, and Hvding 1993; King and Fa'asili 1997; Ruddle 1994). "Co-management is defined as the sharing of responsibility and authority between the government and local fishers/community to manage a fishery or other natural resources" (Pomeroy and Williams 1994). In both Samoa (King and Fa'asili 1997) and Okinawa (Kakuma and Higa 1995), the co-management of inshore resources has been implemented with fisheries extension programmes in analogous ways.

State of the resources

In the Pacific Islands, bottom fish and sedentary species such as groupers, snappers, black pearl oyster, giant clams, trochus, and *bêche-de-mer* (sea-cucumber) have been reportedly overharvested. However, it is difficult to confirm this statistically, even in Fiji – which has the most sophisticated fisheries statistics in the South Pacific countries. In Okinawa, the inshore stocks also have decreased. Fisheries statistics from the Okinawa Development Agency show that catches of bottom fish have decreased drastically since 1982, whereas, for fishing with fish aggregating devices (FAD) and for giant oceanic squid (*Thysanoteuthis rhombus*) fishery, production has increased (fig. 13.1). Production of sedentary species, such as shellfish and sea urchin, has also decreased (fig. 13.2).

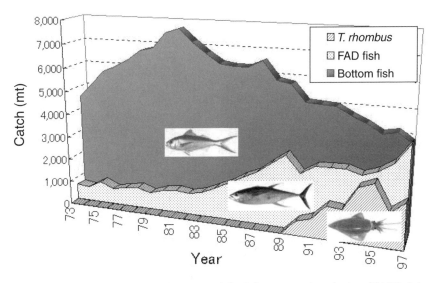

Figure 13.1 Catch transition of bottom fish, fish aggregating device (FAD) fish, and *Thysanoteuthis rhombus* (giant squid). Bottom fish: snappers, emperors, groupers, parrotfish, rabbitfish, fusiliers, and jacks. FAD fish: yellowfin tuna, bigeye tuna, skipjack, dolphin fish, wahoo, blue marlin, and albacore. mt, metric tonnes

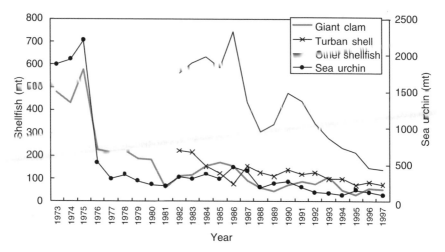

Figure 13.2 Catch transition of shellfish and sea urchin. Giant clam: mostly *Tridacna crocea*; turban shell: *Turbo argyrostomus*; other shellfish: trochus, strawberry conch, etc.; sea urchin: *Tripneustes gratilla*. mt, metric tonnes

The fisheries resources, in essence, need proper management because (a) the resources are renewable, and (b) the resources are considered to be some sort of common property. Unlike fossil resources (e.g. oil or coal), renewable resources (e.g. forest or fish) can be used in a sustainable manner under the proper management, and the growth rates of fisheries resources are usually very high. Although fish in the wild are considered to belong to society in general, it is individual fishermen who utilize these resources. They may strive to obtain as many fish as possible from the resources, even though there must be an optimum harvest level for the society as a whole. The result sometimes, therefore, is the undesirable situation of overfishing. In open-access fisheries, a 60 per cent reduction in effort (e.g. in the number of fishing outfits) would most benefit the fisheries (McManus 1996); this is the so-called "tragedy of the commons."

Conditions in tropical/subtropical fisheries

Tropical and subtropical fisheries have the following characteristics:
1. *Many more species than temperate regions*: In the temperate countries – Japan, for example – the top ten species together accounted for 40 per cent of the total production value in 1996. In contrast, 200–300 species are usually harvested in the typical tropical islands (Munro

1996). In Okinawa, the combined production value of the species of which individual percentages are less than 3 per cent accounts for 70 per cent of the total production in 1997 (from the Okinawa Fisheries Experimental Station Statistics). The fisheries management is basically manipulated to target single species: the more the species, the harder the tasks for the management.
2. *Many small islands and remote areas*: "Regulations are almost useless if they cannot be enforced" (Adams 1996). Enforcement or policing is difficult in tropical countries that have numerous remote areas and islands.
3. *Fewer researchers*: Most of the tropical, developing countries have fewer scientific researchers to deal with fisheries management.
4. *Large percentage of small-scale, subsistence fisheries*: The tropical fisheries can be divided into "commercial," "artisanal" (small-scale), and "subsistence" fisheries. The subsistence fisheries' production in the South Pacific covers more than 80 per cent of the total production (Dalzell, Adams, and Polunin 1996). The subsistence fisheries are rarely monitored for statistics, which hinders management.
5. *Stronger sense of community*: In the South Pacific countries and Okinawa, people traditionally have a strong sense of community and more often than not have managed the local resources by community-based methods (Ruddle 1996; Johannes, Ruddle, and Hvding 1993), which differ from the Western government-based management methods.

Approaches and tools for management

Sociological and ecological approaches have recently been required for the inshore-resource management, adding to traditional biological and economic approaches. Adams (1996) observed: "Compliance is always more desirable than enforcement. There is no such thing as a single-species fishery; all fisheries involve at least two species and one of them is *Homo sapiens*." We should consider more the people's side of the management, as it is the people, not the fish, who practise the management. There is a growing tendency for fisheries management to mean not only control of the fish-stock abundance but also preservation of the habitats and ecosystems of coral or mangrove areas.

The inshore resources might be managed more effectively with co-management than centralized management. The merits of co-management are that it is easy to achieve compliance, more effective and cheaper to enforce, able to incorporate the fishermen's knowledge, and flexible. It may not appear sufficiently "scientific" to incorporate fishermen's ideas;

however, this cost-effective information can be analysed scientifically. Sometimes, fishermen know more than the researchers about when and where the target species spawn (one of the most important pieces of information for management). They also know the status of the stocks they target, or to what types of restrictions they can better conform. The demerits of co-management are that the restrictions decided on by the local community do not have firm legal bases, and such restrictions are not effective for the species that migrate beyond the community's management range. The by-law system could alleviate the former demerit (Fa'asili and Kelekolo 1999); for the latter demerit, we probably should introduce such centralized management regimes as the Total Allowable Catch (TAC) system.

There is another contrast in management approaches, between "preemptive" and "retrospective." "Pre-emptive management requires detailed knowledge of the target organisms and fishing community. By contrast, retrospective management is a form of adaptive management in which rules are developed on the basis of experience of their effects on stocks" (Adams 1996). Although either extreme would be unsuitable, the retrospective approach seems better for management in the tropics, because all five conditions of the tropical fisheries (many species, many remote islands, few researchers, many subsistence fisheries, strong sense of community) suggest that we have to select retrospective management. In Okinawa, probably only the management of a species of emperor fish (*Lethrinus atkinsoni*) can be termed pre-emptive management. The length, otolith, age, growth, mortality, and catch statistics of the fish were studied and, after assessing the stock, the government suggested management measures. This study of just one species was time-consuming – and we have at least eight economically important species of emperor (more than fifty, if we count all the important inshore species). Thus, it could take decades to complete research on all the species, and the stocks may have been severely depleted by then. The management of the newly found giant squid (*T. rhombus*; Kawasaki and Kakuma 1998) and co-management in Onna village (Kakuma and Higa 1995) are other types of retrospective management.

For retrospective management, one needs to consider the following:
1. As the information for the management might not be enough, there is a risk in the result of the management. The approach should be cautious (Food and Agriculture Organization [FAO] 1995, 1996);
2. The results, especially those from the fisheries, should be fed back so that measures can be modified effectively and the management plans should be designed accordingly (Adams 1996);
3. The management plans should be flexible, to be modified efficiently.

Among the management tools (seasonal closure, size limit, tackle restriction, catch quota, licence, etc.), that of a marine reserve seems most effective in the tropics (Bohnsack 1996; King and Fa'asili 1999; Munro 1996). This could also benefit the preservation of coral or mangrove areas. The idea of "source and sink" should be clearly organized. The larvae, juveniles, or adults of the target species could spread out of the reserves (sources) and settle outside (i.e. in the fishing grounds or "sinks.") The size and spacing of the reserves will be determined accordingly. To date, Okinawa has two legal fish reserves, Nagura and Kabira in Ishigaki Island. These were designated by the Ministry of Agriculture, Forestry, and Fishery and a great many procedures would be necessary to modify the restrictions surrounding them. This shows that complicated procedures are inappropriate for community-owned small-scale reserves; the regulations of such reserves should be flexible.

A comparative case study between Samoa and Okinawa

In Samoa, 44 fishing villages implemented coastal-resources management, including protection of critical habitats such as mangroves under a community-based fisheries extension programme. Many (38) of them chose to establish small village fish reserves in their traditional fishing areas. The prospects for continuing compliance and commitment to the management by the village people appear to be good (King and Fa'asili 1999). Figure 13.3 shows the extension process in Samoa.

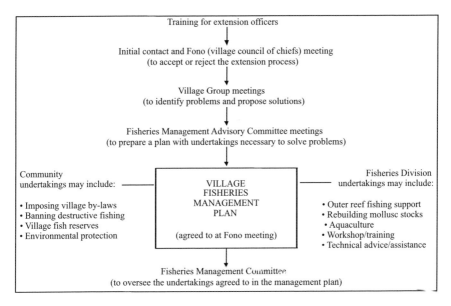

Figure 13.3 The fisheries extension process in Samoan villages. Source: King and Fa'asili (1999)

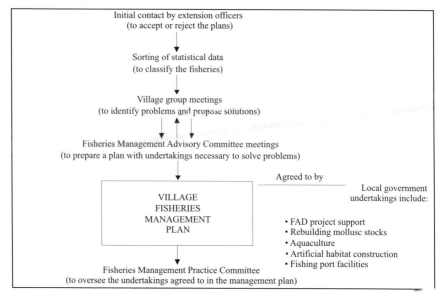

Figure 13.4 The fisheries extension process in Onna Village, Okinawa

In Okinawa, the Onna village fisheries cooperative has practised sedentary-resource management ever since a fisheries extension programme was introduced there in 1988. The target species were trochus, giant clams, turban shell, strawberry conch, sea urchin, spiny lobster, and damselfish. Of these, a giant clam (*Tridacna crocea*) is most important and the management was forwarded with re-stocking of hatchery-produced juveniles. Figure 13.4 shows the extension process. The management involved establishment of marine reserves (fig. 13.5). The process is quite similar to that in Samoa, especially the extensive involvement of the communities in management, in both cases. The governments mainly provided scientific information and rarely compelled the fishermen to restrict the harvest. Both types of management have seemed successful to date, and we might expect them to spread to other fishing villages in tropical/subtropical regions.

Alternative sources of income for the fishermen

Alternative income for the fishermen is essential because, in the management plan, sometimes initial reduction of the harvest is required. In the Samoan case it was thought that (a) the introduction of medium-sized

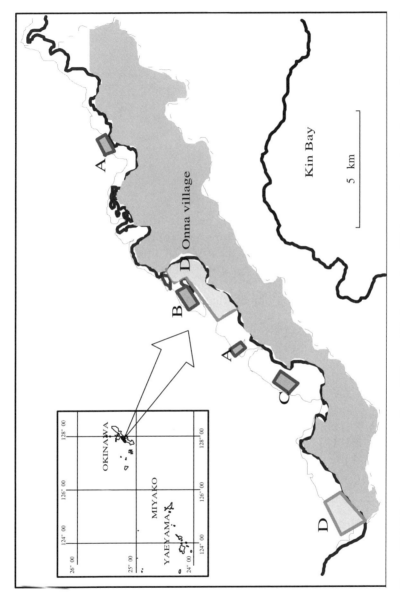

Figure 13.5 Marine reserves in Onna Village: (A) for trochus; (B) for trochus, giant clam, and turban shell; (C) for giant clam; (D) for sea urchin

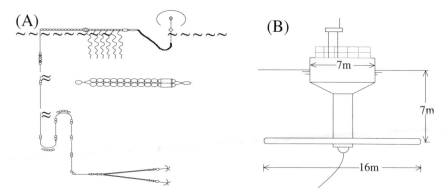

Figure 13.6 Configurations of Okinawan FADs: (A) cheapest type; (B) most expensive type

low-cost boats to diversify the fishing pressure to outer reef areas, (b) the promotion of village-level aquaculture, and (c) the re-introduction of depleted species would be suitable measures. Although aquaculture might seem the simplest remedy, it is controversial in terms of the environment, especially concerning mangrove protection.

FADs (artificial floating objects that attract many pelagic fish, such as tuna and skipjack) may play an important role as the alternatives. In Okinawa, about 150 FADs were deployed recently and produced 2,500–4,000 t of fish annually (Kakuma 1999). FADs are now popular worldwide. In the Philippines (the origin of FADs), there are about 2,000 FADs, producing more than 300,000 t tuna (Dickson and Natividad 1999). FADs are important for commercial fisheries, such as large-scale purse seine, but also for small-scale artisanal fisheries. There are a variety of types of FAD in Okinawa that cost from $4,000 to more than a million dollars. Figure 13.6 depicts, diagrammatically, the configurations of the cheapest and the most expensive types. There are cheaper ($550) FADs in Caribbean waters (Taquet, Gervain, and Lagin 1998).

Gill-net fishery in Iriomote Island

Recent annual fisheries production in Iriomote Island (which is home to the majority of Okinawan mangroves) is 50–55 t; there are about 50 fishermen. Gill-net fishery is most popular in mangrove areas. A rabbitfish, golden-lined spinefoot (*Siganus canaliculatus*), is the most abundant of a variety of fish species caught by gill-net. Table 13.1 shows the major species caught in the Taketomi district (including Iriomote Island and

Table 13.1 Major fish species caught in Taketomi District (kg)

English name	Scientific name	1996	1997	1998	Average
Pearl-spotted spinefoot	*Siganus canaliculatus*	18,246	13,466	18,690	16,801
Octopus	*Octopus cyaneus*	8,339	6,344	5,784	6,822
Cuttlefish	*Sepia latimanus*	5,013	7,515	6,110	6,213
Yellow-tailed emperor	*Lethrinus atkinsoni*	6,095	5,650	5,992	5,912
Squid	*Sepioteuthis lessoniana*	3,660	2,338	2,533	2,844
Rudder fish	*Kyphosus cinerascens*	2,680	3,651	1,573	2,635
Porcupine fish	*Didon liturosus*	2,972	2,280	2,592	2,615
Thread-fin emperor	*Lethrinus nematacanthus*	3,110	2,461	2,134	2,568
Spangled emperor	*Lethrinus nebulosus*	2,852	2,377	1,745	2,325
Golden-lined spinefoot	*Siganus guttatus*	1,422	1,868	1,462	1,584

nearby small islands) in 1996, 1997, and 1998. The majority of the catch was by gill-netting, but the catch includes those caught by small-scale set net and by hook and line in coral areas. According to the fishermen, the fisheries resources have been decreasing and some kind of management is needed; however, the high cost of transport of the catch is the major problem for the fisheries.

REFERENCES

Adams, T.J.H. 1996. "Modern institutional framework for reef fisheries management." In: N.V.C. Polunin and C.M. Robert (eds). *Reef Fisheries*. London: Chapman and Hall, 337–360.

Bohnsack, J.A. 1996. "Maintenance and recovery of reef fish productivity." In: N.V.C. Polunin and C.M. Robert (eds). *Reef Fisheries*. London: Chapman and Hall, 283–314.

Dalzell, P., T.J.H. Adams, and N.V.C. Polunin. 1996. "Coastal fisheries in the Pacific Islands." *Oceanography and Marine Biology: An Annual Review 1996*. Guildford: UCL Press, 34: 395–531.

Dickson, J.O. and A.C. Natividad. 1999. "Tuna fishing and a review of 'Payaw' in the Philippines." Presented at the symposium, *Tuna Fishing and Fish Aggregating Devices (FADs)*, Martinique, October 1999. Plouzane, France: Ifremer, 20 pp.

Fa'asili, U. and L. Kelekolo. 1999. "The use of village by-laws in marine conservation and fisheries management." *SPC Traditional Marine Resource Management and Knowledge. Information Bulletin No. 11*, September 1999, 7–10.

Flores, E.C. and C.L. Marte. 1996. *The possibilities of technical cooperation between Southeast Asia and Okinawa Prefecture in Japan. Part 2: Fisheries*. Joint Research Program Series No. 117. Tokyo: Institute of Developing Economies, 65–114.

Food and Agriculture Organization. 1995. *Code of Conduct for Responsible Fisheries*. Rome: FAO, 41 pp.

Food and Agriculture Organization. 1996. "Precautionary approach to capture fisheries and species introductions." *FAO Technical Guidelines for Responsible Fisheries Vol. 2*, 54. Rome: FAO.

Johannes, R.E. and M. Riepen. 1995. "Environmental, economic, and social implications of the live reef fish trade in Asia and the Western Pacific." 82 pp.

Johannes, R.E., K. Ruddle, and E. Hvding. 1993. "The value today of traditional management and knowledge of coastal marine resources in Oceania." *Workshop on People, Society and Pacific Island Fisheries Development and Management. Selected Papers*. Noumea, New Caledonia: SPC. 1–7.

Kakuma, S. 1999. "Synthesis on moored FADs in the North West Pacific region." Presented at the symposium, *Tuna Fishing and Fish Aggregating Devices (FADs)*, Martinique, October 1999, 16 pp.

Kakuma, S. and Y. Higa. 1995. "Sedentary resource management in Onna village, Okinawa, Japan." *SPC and FFA workshop on the management of South Pacific inshore resource fisheries, Manuscript collection of country statement and background papers, Vol. 1*, 427–438.

Kawasaki, K. and S. Kakuma. 1998. "Biology and fishery of *Thysanoteuthis rhombus* in the waters around Okinawa, southern Japan." *International Symposium on Large Pelagic Squids, Japan Marine Fishery Resources Research Center, Tokyo*, 183–189.

King, M. and U. Fa'asili. 1997. "Community-based management of fisheries and marine environment." *Fisheries Management and Ecology*, 6: 133–144.

King, M. and U. Fa'asili. 1999. "A new network of small, community-owned village fish reserves in Samoa." *SPC Traditional Marine Resource Management and Knowledge Information Bulletin No. 11*, September 1999, 2–6.

Ledua, E. and V. Vuki. 1997. "The inshore fisheries resources of Fiji." Presented in *Pacific Science Inter-congress, 7–12 July, 1997*. Suva, Fiji: USP, 9 pp.

McManus, J.W. 1996. "Social and economic aspects of reef fisheries and their management." In: N.V.C. Polunin and C.M. Robert (eds). *Reef Fisheries*. London: Chapman and Hall, 249–282.

Munro, J.L. 1996. "The scope of tropical reef fisheries and their management." In: N.V.C. Polunin and C.M. Robert (eds). *Reef Fisheries*. London: Chapman and Hall, 1–14.

Pomeroy, R.S. and M.J. Williams. 1994. *Fisheries Co-management and Small-scale Fisheries: A Policy Brief*. Manila, Philippines: International Center for Living Aquatic Resources Management. 15 pp.

Ruddle, K. 1994. *A Guide to the Literature on Traditional Community-based Management in the South Pacific*. FAO fisheries circular No. 869. Rome: FAO, 114 pp.

Ruddle, K. 1996. "Traditional management of reef fishing." In: N.V.C. Polunin and C.M. Robert (eds). *Reef Fisheries*. London: Chapman and Hall, 315–336.

Taquet, M., P. Gervain, and A. Lagin. 1998. "Recovering FADs lost at a depth of 2000 m." *SPC FAD Information Bulletin No. 2*, 30–35.

United Nations. 1997. *Sustaining Livelihoods, Promoting Informal Sector Growth in Pacific Island Countries*. Suva, Fiji: UNDP, 52–60.

14

Mangrove Rehabilitation and Coastal Resource Management Project of Mabini–Candijay, Bohol, Philippines: Cogtong Bay

Robert S. Pomeroy and Brenda M. Katon

Introduction

Co-management can be defined as a partnership arrangement in which government, the community of local resource users (fishers), external agents (non-governmental organizations [NGOs] and academic and research institutions), and other fisheries and coastal-resource stakeholders (boat owners, fish traders, moneylenders, tourism establishments, etc.) share the responsibility and authority for decision-making over the management of a fishery. The partners develop an agreement that specifies their roles, responsibilities, and rights in management. Co-management covers various partnership arrangements and degrees of power sharing and integration of local (informal, traditional, customary) and centralized government management systems. There is a hierarchy of co-management arrangements, from those in which the fishers are consulted by the government before regulations are introduced to those in which the fishers design, implement, and enforce laws and regulations with advice from the government. It is generally acknowledged that not all responsibility and authority should be vested to the local level. The amount of responsibility and/or authority that the state level and various local levels have will differ and depend upon country and site-specific conditions. Determining what kind of, and how much, responsibility and/or authority should be allocated to the local levels is a political decision.

Cogtong Bay is an important example of a co-management arrangement. The Mangrove Rehabilitation and Coastal-Resource Management Project (MRCRMP) as a whole is a coastal-resources management project, not just a fisheries-specific project. Integral to the project goal has been the issuance of Certificate of Stewardship Contracts (CSC) to provide tenurial security to mangrove growers. However, the ultimate goal of most project cooperators is linked to a healthy and abundant fishery. The situation provides an excellent opportunity to determine whether co-management is a management option suitable only to single-resource system management (i.e. fishery) or if it can provide broader coastal-resources management.

Since the project addressed coastal-resource management, a more holistic approach to analysis was called for. Thus, both the mangroves and the fishery were assessed. An analysis of mangroves and mangrove users is necessary because the establishment of a co-management arrangement partly hinged on the issuance of CSCs. An analysis of the fishers and the fishery is, likewise, important because a healthy fishery is the goal of local project implementers. In Cogtong Bay, almost all mangrove cooperators in the project are fishers; however, not all fishers were involved in replanting the mangroves.

Cogtong Bay

Cogtong bay is located in the Central Visayas region of the Philippines. Two municipalities, Mabini to the north and Candijay to the south, share the bay's 10,000 hectares of municipal waters (fig. 14.1). Limestone hills and a thin fringe of mangroves are found at the outer portions of the bay. The inner portion has extensive mangrove stands bordered by rice fields and coconut lands. Of 2,000 ha of mangrove forest, 1,400 ha are still intact. Of these, about 275 ha in the islands of Lumislis, Kati-il, Tabondio, and Calanggaman were declared as mangrove wilderness by the national government. These are characterized by secondary bushy growth, having been felled repeatedly in the past. The rest of the mangrove areas, comprising about 600 ha, have been converted to fish-ponds.

The coastal villages of Cogtong in Candijay and Marcelo in Mabini are inhabited by native Boholanos (83 per cent) and other Visayans from neigbouring provinces. The predominant religion is Christian (Roman Catholic). The village residents are fairly homogeneous in terms of ethnicity, religion, and occupation. About three-quarters of the population at the case-study sites rely on coastal resources for survival and livelihood, indicating a high degree of dependence on coastal resources. As well as fish, most families gather crabs, shellfish, algae and other marine products for subsistence, as well as for sale to local markets.

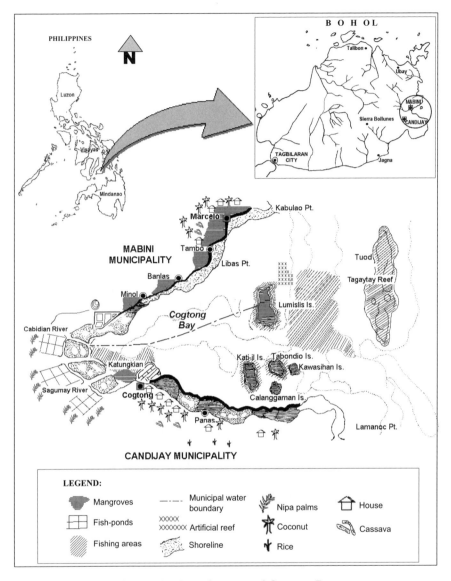

Figure 14.1 Location map of Cogtong Bay

The population of Cogtong has grown from 2,508 persons (434 households) in 1988 (pre-project year) to about 3,361 persons (or 561 households) in 1997. This represents a population increase of 34 per cent since 1988 and an increase of 29 per cent in the number of households during the same period. Mabini's population has grown more slowly from 675 persons (or 120 households) in 1988 to about 777 persons (or 144 households) in 1997. This translates to an increase of 15 per cent over 1988, or approximately 1.7 per cent annually.

The Cogtong Bay fishery is multi-species, multi-tackle, and mainly artisanal. Fishing operations are generally carried out with small, non-motorized boats in dispersed fishing grounds, both outside and within the bay. Small pelagic species, including sardines and mackerel, are caught offshore. Rabbitfish, mullet, trevally, wrasse, scad, and snapper are caught within the bay. The types of fishing tackle used are gill-nets, handlines, fish corrals, spears, fish traps, squid jiggers, and Danish seine.

Brief history of resource management at the site

Historically, Cogtong Bay has been marked by open access, where unrestricted entry to the waters and free-for-all harvesting of coastal products prevailed until the mid-1980s. The bay has no customary rights of tenure to the fishery; for the mangrove areas, however, some form of informal management and tenurial rights has existed for three generations of residents in Cogtong, Candijay. Some 25 families informally took care of small mangrove areas of one hectare or less per family; informal tenurial rights were passed on to the succeeding generations. Eventually, these rights became formal when the third generation applied for mangrove stewardship contracts in the latter half of the 1980s.

Fishers in Cogtong Bay recalled that fishery resources were abundant and mangrove stands were dense until the 1960s. Resource abundance, together with the use of non-destructive harvesting practices and the predominance of subsistence village economies, enabled the coastal residents to utilize coastal resources without major conflicts in resource use. The next two decades, however, saw a drastic change in the situation due to three major developments: these include the introduction of fish-pond technology from Iloilo, a province in the Western Visayas region; the arrival of commercial fishers and entry of commercial mangrove cutters from neighbouring provinces; and the integration of Cogtong Bay into the heavily market-driven economies of nearby provinces and urban centres, such as Cebu and Tagbilaran. These factors hastened the degradation of the bay's resources and resulted in conflicts among resource users. The open-access resource and lack of vigilant law-enforcement efforts fostered the use of illegal fishing practices (i.e. the use of fine-mesh nets and blast fishing) as well as rampant mangrove cutting for firewood and for fish-pond development. The situation was aggravated by fragmented resource-management functions among national government agencies and a lack of leadership, which gave rise to unclear jurisdiction over coastal-resource management. The shift from subsistence village economies to market-driven economies opened new linkages to provincial and regional markets in the Visayas, which intensified resource use.

The devastation of mangroves and fisheries posed a serious resource problem and was a source of discontent among coastal residents, whose very survival is intertwined with the bay's resources. Village fishers became increasingly aware of the decline in their average fish catch over time: their average catch reportedly dwindled from about 20 kilos in the 1960s, to 10 kilos in the 1970s, to approximately 5–7 kilos in the 1980s. The native residents found disturbing the influx of non-coastal residents and outsiders from neighbouring provinces, who destroyed mangrove areas to make fish-ponds.

Changes in resource management: Major initiatives

In 1989, a major effort to avert resource degradation in Cogtong Bay and to promote more sustainable coastal-resource management came about through the initiative of ACIPHIL, Inc., an externally based private firm that has actively provided technical assistance to resource-management projects in the Philippines, including the Central Visayas Regional Project. ACIPHIL entered into a partnership with the Department of Environment and Natural Resources (DENR) of the national government to pursue mangrove rehabilitation and coastal-resource management as a component of the USAID-funded Rainfed Resources Development Project (RRDP). Inspired by the inshore fisheries component of the World Bank-assisted Central Visayas Regional Project (1984–1992), the Cogtong Bay project of Mabini–Candijay sought to transform resource users into resource managers who are directly responsible for day-to-day resource decisions. It adopted a co-management approach to address the problem of resource degradation and poverty in coastal villages along Cogtong Bay from 1989 to 1991 (Janiola 1996; Mehra, Alcott, and Baling 1993; Network Foundation 1990).

The project featured a set of interventions and a process of empowering coastal villagers to carry out their own development and manage their renewable resources. Community organizers were hired as catalysts to initiate awareness campaigns, strengthen local capabilities, forge linkages with government units, and establish village-based fishers' associations for coastal-resource management. In line with efforts to improve the condition of coastal resources, the project introduced mangrove management as a major intervention. Complementing mangrove management were other project components, such as community organizing, capability building, environmental education, mariculture (i.e. culture of mussels and oysters), concrete artificial reefs, and project facilities. The Network Foundation, an NGO, assisted ACIPHIL in implementing the MRCRMP of Mabini–Candijay between 1989 and 1991. Among the project's physical accomplishments were the organization of 13 fishers' associations, issuance of 265 CSCs for mangrove tenurial rights, the rehabilitation of 110 ha of mangrove areas in Mabini and Candijay, and the installation of 265 modules of concrete artificial reefs.

The assistance of the government was visible in several ways, including the provision of funds, resolution of policy issues, law enforcement, passage of enabling legislation, and issuance of property rights. Central to project implementation was the provision of secure mangrove tenurial rights to local fishers. The MRCRMP phase (1989–1991) ushered in the redefinition of access to mangrove areas and the establishment of formal tenurial rights through the issuance of 25-year CSCs. The DENR gave CSC holders the right to manage their mangrove areas and harvest their trees, provided that they replanted each tree felled. Non-CSC holders were not allowed to fell mangrove trees in CSC-covered areas. This period also saw the need for a clearer delineation of political and legal boundaries to address issues of jurisdiction and resource use. The fragmentation of functions for coastal-resource management at that time was manifested in the jurisdiction over mangrove areas by the DENR and in the authority of the Bureau of Fisheries and Aquatic Resources (BFAR) over fisheries (DENR et al. 1997).

During project implementation, a closer coordination between the DENR and the BFAR became imperative to resolve conflicting policies on resource use and fish-pond development. BFAR at that time was encouraging fish-pond development and issuing Fishpond Lease Agreements (FLAs). In some instances, this led to the clearing of well-stocked mangrove forests for fish-pond construction. Village residents asked why they were expected to plant new mangroves and refrain from felling existing trees when outsiders were allowed to come in and destroy mangrove forests. The struggle between FLA holders and village fishers was resolved when the DENR ruled that felling trees in mangrove forests for fish-pond development was illegal. In the absence of felling permits from the DENR, FLA holders could not fell mangrove trees legally.

Recognizing the importance of strict and vigilant law-enforcement efforts, the project staff and village fishers' associations linked up with the municipal government of Mabini and Candijay for support in terms of facilities, police officers, and local legislation. The management of Cogtong Bay's resources called for a committed partnership between the government and the village residents. Joint patrol teams regularly guarded their coastal waters and mangrove areas. Although prevention of illegal fish-ponds was not envisaged as a project activity, the fishers' associations felt that the problem was serious enough to warrant collective action. In many instances, they succeeded in preventing the construction of illegal fish-ponds and the illegal harvesting of mangroves for commercial sale; they also played an active role in controlling blast fishing in the bay.

During the post-MRCRMP phase, however, fishers observed a lower level of rule compliance. This was due, in part, to weaker law enforcement and less support from the municipal government that came with a change in political leadership and with budgetary constraints. Consequently, the lack of vigilance and the breakdown in enforcement efforts encouraged illegal fishers to resume their destructive activities in Cogtong Bay. Illegal mangrove felling, however, was less problematic in areas with formal property rights: the CSC holders, on their own, continued to protect their mangrove areas.

Political boundaries became more distinct when the Local Government Code effected the devolution to local government units of many of the functions previously performed by BFAR and DENR. At present, the municipal government exercises jurisdiction over municipal waters (i.e. waters within 15 km of the shoreline of the municipality) and over the management of community-based forestry projects. Areas beyond the municipal waters, as well as those outside communal forests, however, remain under the BFAR and DENR, respectively.

In recent years, the Village and Municipal councils of Candijay and Mabini have demonstrated a stronger interest in coastal-resource management. They have supported the establishment of a new fish sanctuary at Lumislis Island, pushed for stricter local legislation, and recognized communal mangrove areas for firewood gatherers.

Incentives to cooperate

The shift from open access to a communal property-rights regime for mangrove areas in Cogtong Bay was prompted by several factors. These include: (1) a common dependence on coastal resources on the part of resource users; (2) a desire for better coastal-resource management on the part of government organizations and NGOs; (3) concern for improving the socio-economic condition of poor coastal residents; (4) the legitimacy of property rights; and (5) realization of the need for collective action against illegal fishing and illegal mangrove felling to avert further resource degradation.

Disincentives to cooperate, on the other hand, initially stemmed from conflicting government policies and the indifference of some local government officials to strict law enforcement. These were eventually resolved when the MRCRMP drew attention to these areas and, together with fishers' associations, put pressure on appropriate organizations to take action.

The delineation of mangrove property rights, however, led to the displacement of firewood gatherers. The restriction of harvesting rights to CSC holders alienated the firewood gatherers, but this was resolved with the designation of communal mangrove areas for firewood gathering. To prevent the rapid depletion of mangroves, the municipal government passed a local ordinance that banned the sale of mangrove firewood outside Mabini and Candijay. The intent was to meet the needs only of the domestic market.

Outcomes of co-management arrangements

The co-management experience in rehabilitating mangroves at Cogtong Bay may be regarded as a victory over indiscriminate mangrove felling on a large scale and a demonstration of a greater consciousness among resource users of the interaction between humans and the environment. Nurturing the resource base and protecting livelihoods over the longer term called for a shift from a "use orientation" to a "resource-management orientation" that actively seeks the enhancement of knowledge, skills, responsibility, and accountability of resource users and other stakeholders. Central to this perspective is the recognition that resource users are de facto managers of the natural resource base and stewards of the environment in which they live.

Mangrove rehabilitation and the provision of stewardship contracts to mangrove planters have paid off. A post-project mangrove assessment in July 1997 indicates relatively good mangrove growth at the reforested areas – particularly at Katungkian, which seems to have been influenced by protection from waves, relatively shallow depth, extensive water runoffs, and a muddy soil layer. The total basal area of mangrove stands was highest in Katungkian, at 6.82 m^2/ha; it was slightly lower at other sites, such as Panas, Catiil, and Lumislis. Compared with San Miguel Bay, another site in the Philippines where overharvesting and mangrove conversion to aquaculture has also led to mangrove denudation, mangrove growth at Cogtong Bay is relatively better (Vega et al. 1995).

An assessment of coral-reef conditions in Cogtong Bay, using manta-tow reconnaissance survey, indicates a percentage live-coral cover of 11–50 per cent (poor to fair) on the eastern side of Lumislis Island; Tagaytay reef, which lies near the mouth of the bay, has a higher percentage of live-coral cover at 51–75 per cent (good). In relation to fish catch per fishing trip, key informants have noted a continuing downward trend in the 1990s for various types of fishing tackle, particularly for longlines, fish pots, handlines, spear guns, and squid jiggers. The only exceptions apply to gill-nets and fish corrals, where the average fish catch of about 3–10 kg in 1997 comes close to the estimated catch in the 1980s. Although marine sanctuaries were established in 1995, they are relatively recent. The absence of benchmark data, moreover, precludes conclusions on the extent to which a change in fish abundance and living-coral cover has occurred.

From the perspective of fishers, positive socio-economic changes occurred in a number of indicators, given a comparison of the situation before (1988) and after (1997) the project (Pomeroy et al. 1996). Larger positive changes were perceived in knowledge of mangroves, information exchange on both mangrove management and fisheries management, control over mangrove resources, speed of resolving community conflicts, and influence over community affairs. They also perceived an improvement in household income and household well-being, among others. The fishers, however, did not perceive a statistically significant change in the overall well-being of coastal resources. This perception seems to have been influenced by the relatively lax enforcement of fishery rules during the post-project phase, the return of illegal fishing activities in recent years, and a general downturn in fish catch for simple types of fishing tackle. The fishers, none the less, are optimistic that the fishery will improve in the next five years. It appears that the co-management regime has been more successful in mangrove management owing, in part, to the issuance of stewardship contracts that have provided a strong incentive to protect the mangrove stands and the relative ease in patrolling the mangrove areas on the part of the mangrove planters.

The experience of Cogtong Bay affirms that the management of coastal resources is not easy. It was marked by obstacles at various stages that ranged from fragmentation of functions on coastal-resource management, to conflicting government policies on fish-pond development in the 1980s, to weak law-enforcement efforts after project completion in 1991. It draws attention to the difficulty of managing coastal resources without the sustained cooperation of the government and the resource users to make rules and regulations work, particularly for fisheries management. Gains, none the less, are possible in spite of formidable odds if resource users and the government have a shared commitment to sound resource management and livelihood protection and are willing to take decisive action.

The fairly successful co-management experience of Cogtong Bay may be attributed, in part, to the painstaking efforts of the MRCRMP, the provision of legitimate property rights through issuance of CSCs by the national government, and the deliberate involvement of CSC holders and resource users as active partners in coastal-resource management. The continuing existence of mangrove stands established by the project along Cogtong Bay, the persistence of holders of mangrove stewardship contracts voluntarily to guard their respective mangrove stands from illegal fellers, and the recent emergence of new resource-management initiatives (i.e. marine sanctuary) provide concrete proof that resource stakeholders have made a breakthrough. The project has imbued the village residents with a sense of empowerment and has built their confidence to take action on collective concerns. It has invoked the basic principle of control and accountability, where control over action rests with the people who will bear its consequences. It has also helped village residents to understand their own situation and to address felt needs jointly.

Characteristics of functional co-management arrangements

Given below are insights into the characteristics of functional co-management institutional arrangements in Cogtong Bay, which have been drawn from key informant interviews, regression analysis, household survey data, and secondary data.

1. *Recognition of a resource crisis/resource–management problem.* The Cogtong Bay experience affirms that the recognition of resource-management problems (i.e. declining mangrove stands, lower fish catch, disappearance of high-value fish species, conflict between resource users) prompts resource users to enter into collective arrangements, largely because of the threat to their survival and livelihood.
2. *Extent of dependence on coastal resources.* A heavy dependence on coastal resources places the survival and livelihood of village residents at risk if nothing is done to halt progressive resource degradation.
3. *Capability building.* Co-management requires a conscious effort to develop and strengthen the capability of resource users and stakeholders for collective action, dialogue, leadership, and sustainable resource management.
4. *Trust between partners.* Co-management arrangements are enhanced when mutual trust and respect are in place.
5. *Involvement of resource users in law–enforcement efforts.* Resource users, who have more stake in sustaining their resource base, must be actively involved in monitoring illegal mangrove felling and illegal fishing activities.
6. *Provision and enforcement of legitimate property rights.* In Cogtong Bay, legitimate property rights defined the required mechanisms to optimize resource use and conservation, particularly for mangroves.
7. *Continuing support from local leadership.* During the project-implementation phase, the active partnership between the municipal

government and the village fishers in law enforcement was a potent force in the reduction of illegal fishing and mangrove felling.
8. *Existence of enabling legislation for decentralization.* The decentralization of coastal resource-management functions from the national government to the local government units, primarily through the enactment of the Local Government Code of 1991, helped to clarify jurisdiction over coastal resources.
9. *Shared vision and commitment to sustainable coastal-resource management.* Formal legislation must be complemented by a continuing advocacy of sustainable resource management and by concrete measures to achieve a conscious sharing of responsibility between the government and resource users.

REFERENCES

Department of Environment and Natural Resources (DENR), Department of Interior and Local Government, Department of Agriculture–Bureau of Fisheries and Aquatic Resources, and the Coastal Resource Management Project. 1997. *Legal and Jurisdictional Guidebook for Coastal Resources Management in the Philippines.* Manila: Coastal Resources Management Project.

Janiola, E. 1996. "Mangrove rehabilitation and coastal resources management in Cogtong Bay: addressing mangrove management issues through community participation." In: E.M. Ferrer, L. Polotan-De la Cruz, and M. Agoncillo-Domingo (eds). *Seeds of Hope.* Manila: University of the Philippines, College of Social Work and Community Development.

Katon, B.M., R.S. Pomeroy, M. Ring, and L. Garces. 1998. *Mangrove Rehabilitation and Coastal Resources Management Project of Mabini–Candijay: A Case Study of Fisheries Co-management Arrangements in Cogtong Bay, Philippines. Fisheries Co-management Research Project Working Paper No. 33.* Manila: International Center for Living Aquatic Resources Management.

Mehra, R., M. Alcott, and N. Baling. 1993. *Women's Participation in the Cogtong Bay Mangrove Management Project: A Case Study.* Washington, DC: World Wildlife Fund.

Network Foundation. 1990. *The Cogtong Bay Mangrove Management Project.* Cebu City, Philippines: Network Foundation.

Pomeroy, R.S., R. Pollnac, C. Predo, and B. Katon. 1996. *Impact Evaluation of Community-based Coastal Resource Management Projects in the Philippines. Fisheries Co-management Project Research Report No. 3.* Manila: International Center for Living Aquatic Resources Management.

Vega, M.J.M., L.R. Garces, Q.P. Sia III, and R.G.G. Ledesma. 1995. "Assessment of mangrove resources in San Miguel Bay." In: G.T. Silvestre, C. Lina, and J. Padilla (eds). *Multidisciplinary Assessment of the Fisheries of San Miguel Bay, Philippines (1992–1993). ICLARM Technical Report 47* (in CD-ROM). Manila: International Center for Living Aquatic Resources Management.

Part III
Uses and policies

15

Towards sustainable use and management for mangrove conservation in Viet Nam

Motohiko Kogo and Kiyomi Kogo

Introduction

The extent of the Viet Nam coastline is 3,260 km. The importance of mangroves is of relevance to a great number of inhabitants, probably more than one-half of the national population. Mangroves are distributed along all the coast but large mangrove areas occur only in the deltas of the Mekong and Hong rivers. The forest area of mangroves is 252,500 ha. The number of species is 78, including mangrove associates, with fewer species in the north because of the cold winters (Hong and San 1993).

During the war, mangrove forests in Viet Nam were seriously damaged: about one-half of the area of mangroves in South Viet Nam was destroyed (Baba 1994). Reforestation activities started soon after the unification of North and South Viet Nam in 1976. It is estimated that more than 100,000 ha were reforested without any foreign aid (Kogo and Kogo 1997). However, mangrove forests are still decreasing and degraded, because shrimp farming has expanded with the destruction of mangroves which has taken place along the entire coast from the Chinese to the Cambodian borders, even in plantation mangroves.

Damage to mangrove forests has caused various problems, such as a shortage of firewood and charcoal, reduced marine resources, and increased coastal erosion. Particularly in the north and middle part of Viet Nam, where typhoons are frequent, there has been an increase in loss of human life and damage to property and agriculture caused by flood tides (Izumo and Kogo 1994).

The majority of coastal inhabitants, farmers, and fishermen, have become poorer and poorer. The people welcome mangrove restoration; however, they do not have the means to implement it.

This is a report on international cooperation of non-governmental organizations (NGOs) which support local communities for the reforestation and conservation of mangroves in Viet Nam.

Goal of the project

The project is run by ACTMANG (Action for Mangrove Reforestation), an NGO based in Tokyo. It was started in 1993, following a proposal by Professor Phan Nguyen Hong, Director of the Mangrove Ecosystem Research Division (MERD), Viet Nam National University, Hanoi.

The goal of the project is the restoration of mangrove forests. It is an open-ended project with undetermined closure time. The project is implemented in partnership with MERD and in cooperation with the local communities, the people's committees, and the Women's Union. The cooperation has been successful in all cases except one.

Research

Research was carried out in various fields to understand the situation of Vietnamese mangroves – for example, the utilization of mangroves (Izumo and Kogo 1994), domestic technology for mangrove reforestation (Tsuruda and Asano 1994), international cooperation for mangrove reforestation (Asano 1994), natural conditions for growing mangroves (Miyagi 1995), and life in the village (Suzuki 1999). Three basic research studies are as follows.

Bibliography

According to UNESCO's *Bibliography on Mangrove Research 1600–1976* (1981), there are only 87 scientific papers on Viet Nam's mangroves. However, research by Vietnamese scientists has increased since the end of the 1970s: 211 papers, mainly written in Vietnamese, were added to the list by this research (Hong 1994).

Requirements for reforesting mangroves

Some 500 questionnaires were posted to the people's committees in the whole of the coastal area: the answer rate was 24 per cent at province level and 27 per cent at district level; however, a great deal of information was obtained at local levels, on such topics as the population and industries of the communities, the presence of mangrove areas and their species composition, experience in reforestation, areas in need of reforestation, and others (ACTMANG/MERD 1997). All those who answered stressed their wish to support the project. The reasons for wanting reforestation are as follows:
- *In all coastal areas*: Protection of the land with sea walls; protection against coastal erosion, rehabilitation of ecosystems for increasing fishery resources; supplying firewood and building materials;
- *On the northern and middle coasts*: In addition to the above, protection against flood damage;
- *In specific areas*: Development of eco-tourism is sought.

Function of mangroves in protecting the coast

It is widely recognized that mangroves protect the coast from erosion and damage caused by tropical storms, although there are not many quantitative scientific publications on this topic.

Preliminary field measurements were tentatively undertaken at the muddy coast of Thai Thuy, Thai Binh Province, which is directly exposed to the open sea. The rate of wave reduction per 100 m of mangroves in the direction of wave propagation was measured according to the formula: $r = (HS - HL)/HS$, where r is the rate of wave reduction, and HS and HL are the wave heights at the offshore edge of the mangrove and 100 m inshore in the mangroves, respectively. All measurements were approximate.

On the coast of Thai Thuy, *Kandelia candel* was planted in a strip 1.5 km wide and 3 km long along the coast. Six years after planting, the trees had grown to a height of about 2 m. The wave height of 1 m at the open sea would reduce to 0.05 m at the coast; however, without the sheltering effect of mangroves, the waves would arrive at the coast with a wave height of 0.75 m. For dense mangrove forest, the rate of wave reduction does not decrease with increasing water depth; however, these are tentative and approximate conclusions.

Thus, it appears that mangroves are an efficient protection against erosion of the coast, even at flood tide when the sea level rises substantially (Magi et al. 1996; Mazda et al. 1997). The studies have continued on the coast of Tien Lang (Hai Phong City), Can Gio (Ho Chi Minh City), and Ben Tre in the Mekong delta (Magi et al. 1997; Mazda, Magi, and Kobashi 2000).

Reforestation

An area of 954 ha has been completely reforested in 1994–1999. There are three principal areas – in Thai Thuy, Thai Binh Province (160 ha); Tinh Gio, Thanh Hoa Province (116 ha); and Tien Lang, Hai Phong City (617 ha); all of these are located in the north. In addition, some experiments are still to be implemented in the future, and experiments have already been carried out in Yen Hung, Quang Ninh Province in the north; Hu Loc, Hue in the centre; and Thanh Phu, Ben Tre Province and Vinh Trau, Soc Tran Province in the south (table 15.1, fig. 15.1).

The species for planting were selected on the basis of the natural conditions of the planting site and availability of seeds and seedlings. In the northern areas these were *Bruguiera gymnorrhiza*, *Kandelia candel*, *Rhizophora stylosa*, and *Sonneratia caseolaris*; in central areas, *R. stylosa* was selected; in the south, *R. mucronata* and *S. caseolaris* were chosen.

The survival rate was 10–80 per cent depending on the site conditions and effect of typhoons. The growth rates of the trees planted in northern areas were slower than those in the southern areas owing to lower temperatures. The fast-growing species of *S. caseolaris* deserves special mention: although this species sheds its leaves and stops growing in winter, the tree reached 3–5 m within 4 years of planting.

Plantation was carried out by local farmers, mainly women (photo 15.1).[1] The mean speed of plantation was 10 persons/day/ha for direct sowing of propagules (spacing 1 m × 1 m) and 8.5 persons/day/ha for transplantation of the seedlings (spacing 1 m × 2 m). Nursery practice for *S. caseolaris* was introduced into Viet Nam by the project; however, the women acquired the techniques within two years.

Sustainable management for conservation

When an agreement for cooperation was entered into, the chairperson of the people's committee agreed to conserve the mangroves planted.

The problems of conservation are difficult to solve. In the case of the ACTMANG/MERD Project in Tien Lang, some practical measures have been undertaken. The Women's Union, the local partner, had a strong desire not only for plantation but also for conservation of the mangroves they planted. In this area, it is essential to protect the plantation from

Table 15.1 Reforestation activities of the ACTMANG project in 1994–1999

nmunity	Planted in	Area (ha)[a]	Species[b]
¨huy district, Thai Binh Province: 160 ha			
˹ai	Apr. 1994	102	Kc
˥i	Jul. 1997	1	Rs, Bg
ɔng	Apr. 1995	50	Kc
˥g	Jun. 1995	2	Bg
g	Jul. 1995	5	Rs
ˌstrict, Hai Phong City: 617 ha			
	Apr. 1995	3	Bg
ʒ	Jun. 1995	1	Sc
ʒ	Jul. 1995	5	Rs
Vinh Qua˷ʒ	Apr. 1995	4	Sc
Vinh Quang	Apr. 1996	35	Sc
Vinh Quang	May 1996	23	Sc
Vinh Quang	Mar. 1998	30	Sc
Vinh Quang	May 1998	30	Sc
Vinh Quang	Jul. 1998	20	Sc
Vinh Quang	May 1999	80	Sc
Vinh Quang	Jun. 1999	30	Sc
Dong Hung	May 1996	16	Sc
Dong Hung	Apr. 1997	18	Sc
Dong Hung	Mar. 1998	150	Sc
Dong Hung	May 1998	70	Sc
Dong Hung	Jul. 1998	30	Sc
Dong Hung	May 1999	32	Sc
Dong Hung	Jun. 1999	40	Sc
Tinh Gio district, Thanh Hoa Province: 116 ha			
Binh Ming	Apr. 1999	45	Kc
Binh Ming	Jul. 1999	6	Rs
Xuan Lam	Apr. 1999	23	Kc
Xuan Lam	Jul. 1999	4	Rs
Hai Binh	Apr. 1999	14	Kc
Hai Binh	Jul. 1999	9	Rs
Hai Trau	Jul. 1999	15	Rs
Other areas for experimentation: 61 ha			
Yen Hung District, Quang Ninh Province	1996	10	Sc
Vinh Trau District, Soc Trang Province	1996	6	Rm
Thanh Phu District, Ben Tre Province	1997	44	Sc
Hu Loc District, Hue City	1999	1	Rs

a. Grand total area of the plantation: 954 hectares.
b. Bg, *Bruguiera gymnorrhiza*; Kc, *Kandelia candel*; Rm, *Rhizophora mucronata*; Rs, *R. stylosa*; Sc, *Sonneratia caseolaris*.

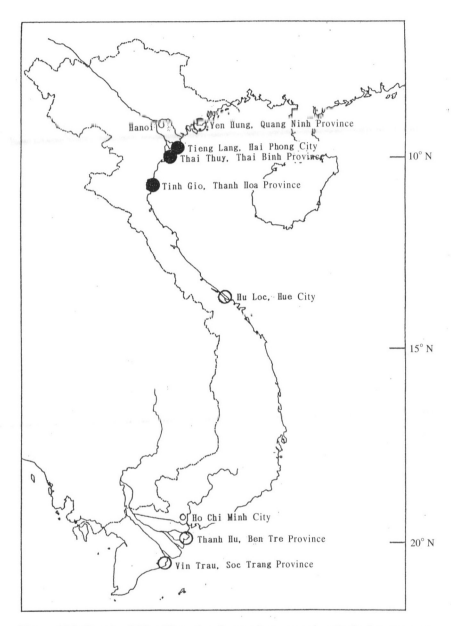

Figure 15.1 Sketch of Viet Nam showing project areas (• principal areas; ○ experimental areas)

cattle browsing, illegal felling, and damage by fishing nets, particularly when trees are young. The Women's Union tried to set a watch in the planting sites on their own initiative. At the same time, they appealed to villagers to cooperate for conservation; thus very little damage was caused by either cattle or humans.

Illegal felling, even in newly planted areas, is carried out by economically poor villagers who can not get enough firewood for cooking (rice straw and coal are used in this area). A vocational training centre was established by the Women's Union. It is also worth mentioning that some percentage of the daily wages that were to be received by individuals participating in planting activities have been deposited for establishing the centre. To enhance the income of the communities, seedling production was also helpful. After the technology transfer for raising seedlings of *S. caseolaris* in a nursery, some 500,000 seedlings were produced by villagers every year and were sold to the project.

It is expected that such activities, carried out by the women and villagers on their own initiative, will establish sustainable management for conservation.

Sustainable use of mangrove ecosystem

Generally speaking, it is strictly forbidden by law to fell mangrove trees, even for thinning. Thus, felling of mangroves is illegal, apart from some cases permitted by the government.

Coexistence of mangroves with shrimp farming

As mentioned, shrimp farming is the greatest problem facing mangrove conservation. Although the situation has not changed, understanding of the importance of mangroves has totally changed because shrimp productivity has decreased year by year, owing to the severe destruction of mangroves. Coexistence between aquaculture and mangroves is strongly sought in Viet Nam today.

A system termed "shrimp farming combined with mangrove reforestation" is widely practised in the Mekong delta area (photo 15.2). This is a system of shrimp farming coexisting with mangroves. The ratio of mangrove area (mainly *Rhizophora apiculata*) to pond area is 7:3 (it was changed to 6:4 later). It has been recognized that mangroves are essential for sustainable shrimp farming. Although excellent in concept, the experiment was not successful. Generally speaking, shrimp productivity is great in the first year but decreases year after year; planted mangroves also grow slowly.

This project was started in 1998, following a proposal by the People's Committee of Ben Tre Province in the Mekong delta. The objective is to improve the system of "shrimp farming combined with mangrove reforestation." The research focused on the comparison between ponds with and without sustainable yield. As a result, the following tendencies were found in those ponds with no sustainable yield (Nishimura 1998, 1999; Mazda, Magi, and Kobashi 2000):
- *Ecologically unsuitable conditions*: sparse soil on the bottom of the ponds, some soil being heavily anoxic owing to the rotting leaves shed by the mangrove trees;
- *Defective structure of the ponds*: higher salinity and lower dissolved oxygen concentrations occur in waters at a distance from the water gates, obviously because of inadequate planning;
- *Insufficient nutrition for shrimps*: scant benthos in the ponds.

Consequently, neither shrimps nor mangroves can grow well.

After completion of the research, three types of improved shrimp ponds (to a total area of 10 ha) were constructed (Miyamoto 1999). However, the result was very poor owing to lack of communication with the local partner: the existing mangroves were totally felled and the method of shrimp culture was not carried out correctly.

Development of eco-tourism

UNESCO suggested that eco-tourism could be a suitable nature-conservation method. The idea of eco-tourism is widely accepted in Viet Nam; however, it is seldom implemented. Can Gio, the southern part of Ho Chi Minh City, is a mangrove area of some 80,000 ha.

During the war, the mangrove forests in Can Gio were completely destroyed by defoliation. Today, almost all the mangroves have been restored by reforestation and conservation through natural regeneration. The reason why this area has been selected is, of course, the history of destruction and restoration of mangrove forests.

The activity was started in 1995 when an agreement with the People's Committee was finalized. The agreement stated that ACTMANG could use 100 ha of mangrove forest for the purpose of education and research over the course of 15 years. During the initial five years, the following activities were undertaken:
1. Research for the development of eco-tourism. For example, research in the mangrove ecosystem, ecology of monkeys, bird watching, boating in narrow creeks, and others.
2. Construction of five lodges (photo 15.3) and a board walk in the mangrove. The lodges have been used by many Vietnamese school

children from Ho Chi Minh City. Japanese scientists and university students also used them for field research.
3. Execution of eco-tourism. Some groups of NGOs, schoolchildren, and girl scouts have been invited from Japan.

When the project was started in 1992, very few people had visited Can Gio. Today the situation is totally changed: a great number of tourists come from Ho Chi Minh City, particularly at weekends. Restaurants and facilities for boating and exploring the mangrove swamps have been established in the mangrove areas by local tourist companies.

In 2000, the mangrove forests in Can Gio were registered as a Biosphere Reserve by UNESCO. However, there is concern that, with the increase of tourism, there could be damage to the mangrove ecosystem; thus, the importance of the project has increased with time.

Education and awareness

Although the importance of education and awareness is widely recognized, very little attention is paid to it in Viet Nam. In this context, it is worth mentioning that the project has received support from many Vietnamese, foreign, and international organizations – for example, universities, women's associations, the Oceanographic Institute of Viet Nam and (from abroad) the Danish Red Cross, OXFAM, Save the Children, Tierra, the International Society for Mangrove Ecosystems (ISME), and other concerns. The main educational activities are shown in table 15.2.

Conclusions

The activities of the project may be roughly divided into two parts – reforestation, and the conservation of mangroves.

Reforestation

A total area of 954 ha has been planted in cooperation with local communities over six years. It was found that farmers (who form a majority in the communities) are extremely able in raising plantations. Undoubtedly, they could complete the reforestation work themselves if financial and technical assistance were provided. Support by NGOs would be effective if a good partnership could be established with the communities.

Although reforestation holds promise, at present there are not enough

Table 15.2 Main activities regarding education

Activities	Year	Contents
1. Production of teaching materials		
• Colour slide set for mangrove education	1995	Produced by ISME/ACTMANG, distributed by MERD to schools
• Picture book on mangrove reforestation	1995	Produced by MERD, expenses of production are donated by foreign NGOs, distributed to farmers living on the coast
• Textbook on mangrove ecosystem	1996	Produced by MERD, expenses of production are donated by foreign NGOs, distributed to schoolchildren
2. Study and training		
• Training courses	1994	Held in field to support reforestation
• Overseas study tour to Indonesia	1995	Supported by ACTMANG/Tierra, for mangrove forest management, 4 officials of HCM City
3. Festivals for awareness		
• Festival of mangrove	1994	In cooperation with Danish Red Cross, plantation held in Thai Thuy
• Music concert	1994	Presented by Mr Shokich Kina, held in Thai Thuy
• Music concert (CAMP98)	1998	Presented by 3 Japanese and 4 Vietnamese groups of musicians, held in Can Gio, joined by 150 Japanese and over 3,000 villagers
4. Workshops		
• 1st Workshop held in Can Gio, HCM City	1994	Organized by MERD/ACTMANG/Agriculture Service in HCM City. Subject: Reforestation and Afforestation of Mangroves in Vietnam; 24 papers presented; proceedings: 187 pp.
• 2nd Workshop held in Do Son, Hai Phong City	1995	Organized by MERD/ACTMANG/Hai Phong Institute of Oceanology. Subject: Reforestation and Management of Mangrove Ecosystems in Viet Nam; 25 papers presented; proceedings: 189 pp.
• 3rd Workshop held in Hue City	1996	Organized by MERD/ACTMANG/Hai Hue University. Subject: Relationship between Mangrove Reforestation and Coastal Aquaculture in Vietnam; 32 papers presented proceedings: 225 pp.

- 4th Workshop held in Hanoi 1997 Organized by MERD/ACTMANG/Vietnam Women's Association. Subject: Socio-economic Status of Women in Coastal Mangrove Areas; Trends to Improve Their Life and Environment; 28 papers presented; proceedings: 197 pp.
- 5th Workshop held in Nha Trang 1998 Organized by MERD/ACTMANG/Institute of Oceanology. Subject: Sustainable and Economically Efficient Utilization of Natural Resources in Mangrove Ecosystems; 35 papers presented; proceedings: 204 pp.
- 6th Workshop held in Hanoi 1999 Organized by MERD/ACTMANG. Subject: Management and Sustainable Use of Natural Resources and Environment in Coastal Wetland; 51 papers presented; proceedings: 307 pp.

areas to be reforested because shrimp farming has completely occupied the coastal zone. It therefore seems advisable to rehabilitate to mangrove ecosystems those shrimp-ponds that do not give a sustainable yield, particularly in areas where erosion and flood tides cause severe damage. The usefulness of mangroves for coastal protection has been demonstrated all over the world, and again by this project. Furthermore, the cost of reforestation is less than the cost of repairing sea walls and paddy-fields.

Conservation

Conservation is a more difficult subject, because many factors lead to the destruction of mangroves – such as illegal felling, conversion to shrimp-ponds and paddy-fields, and others. One ideal measure would be to re-introduce a system of common land ownership, such as the communities had in the old days, but this is not easy to realize because of the great change in social systems.

Sustainable management for the conservation of mangroves by replanting could be realized when local communities participated in this action on their own initiative. The activities of the Women's Union in Tien Lang – such as setting a watch, eliciting awareness by the villagers, and establishing a vocational training centre – are examples of successful initiatives.

Among the activities in education and awareness, it was considered a priority to produce teaching materials for children (who will play an important role in the future) and for farmers (who conserve mangroves directly in their areas). The six workshops that were convened were useful because the participants – not only scientists but also decision makers and farmers – were invited from the areas where mangroves occur. Undoubtedly, the importance of mangroves has been brought home to them by the presentations and by discussions between themselves.

Establishing sustainable shrimp farming with mangroves, and developing eco-tourism, are alternative methods for sustainable use of the mangrove ecosystem. These have not yet been fully tried, but results are expected soon.

Today in Viet Nam, as well as strong promotion of a policy of economic development, 10 million hectares of reforestation (including mangroves but mainly of other tree species) is being undertaken.

Note

1. Throughout this volume, photographs have been placed at the end of the chapter to which they refer.

REFERENCES

ACTMANG/MERD. 1997. *Report of the Research on Communities' Requirement for Reforesting Mangroves in Vietnam* (in Japanese). Tokyo: Japan International Forestry Promotion and Cooperation Center (JFPRO), 92 pp.

Asano, T. 1994. "International cooperation of mangrove reforestation in Vietnam" (in Japanese). In: M. Kogo (ed.). *Report of Basic Research for Cooperating Mangrove Reforestation in Vietnam.* Tokyo: JFPRO, 92–104.

Baba, S. 1994. "Destruction of mangrove forests – the effects by the Defoliation Operation during the Vietnamese war" (in Japanese). In: M. Kogo (ed.). *Report of Basic Research for Cooperating Mangrove Reforestation in Vietnam.* Tokyo: JFPRO, 92–104.

Hong, P.G. 1994. "Bibliography of the research on mangrove ecosystems in Vietnam." In: M. Kogo (ed.). *Report of Basic Research for Cooperating Mangrove Reforestation in Vietnam.* Tokyo: JFPRO, 113–138.

Hong, P.G. and H.T. Sang. 1993. *Mangroves of Vietnam.* Hanoi: IUCN, 173 pp.

Izumo, K. and K. Kogo. 1994. "Utilization of mangrove ecosystems in Vietnam" (in Japanese). In: M. Kogo (ed.). *Report of Basic Research for Cooperating Mangrove Reforestation in Vietnam.* Tokyo: JFPRO, 40–59.

Kogo, K. and M. Kogo. 1997. "Mangrove reforestation in Vietnam by supporting villagers activities – A trial to make a reforestation model on global level" (in Japanese, abstract in English). *Tropics, The Japan Society of Tropical Ecology* 6(3): 247–282.

Magi, M., Y. Mazda, M. Kogo, and P.H. Hong. 1996. "Protection effect against wave action by mangrove reforestation in Tong King Bay delta in Vietnam" (in Japanese). *Bulletin of Faculty of Oceanography, Tokai University, Shizuoka* 41: 157–169.

Magi, M., Y. Mazda, M. Kogo, and P.H. Hong. 1997. "Protective function of planted mangroves in shrimp ponds – a case study in Thuy Hai coast in Vietnam" (in Japanese). In: Y. Mazda (ed.). *Physical Process and Environmental Formation in Mangrove Waters.* Shizuoka: Kurohune Shuppan, 179–185.

Mazda, Y., M. Magi, and D. Kobashi. 2000. *Report of the Research on Physical and Environmental Conditions in Mangrove Waters in Vietnam* (in Japanese). Faculty of Oceanography. Tokai University, Shizuoka, 80 pp.

Mazda, Y., M. Magi, M. Kogo, and P.H. Hong. 1997. "Mangroves as a coastal protection from waves in the Tong King delta, Vietnam" (in English). *Mangrove and Salt Marshes* 1: 127–135.

Miyagi, T. 1995. "Environmental conditions for mangroves in Vietnam" (in Japanese). In: *The Collection Thesis of Tohoku Gakuin University (History and Geography), Sendai*, No. 26, 8–39.

Miyamoto, C. 1999. "Plan of improved shrimp pond for sustainable farming" (in Japanese). In: M. Kogo (ed.). *Report of 2nd Research for Project Formation of Sustainable Shrimp Farming in Vietnam.* Tokyo: JFPRO, 28–35.

Nishimura, K. 1998. "Structure of the pond of 'shrimp farming combined with mangrove reforestation'" (in Japanese). In: M. Kogo (ed.). *Report of Research for Project Formation of Sustainable Shrimp Farming in Vietnam.* Tokyo: JFPRO, 17–35.

Nishimura, K. 1999. "Analysis of 'shrimp farming combined with mangrove reforestation'" (in Japanese). In: *Report of 2nd Research for Project Formation of Sustainable Shrimp Farming in Vietnam*. Tokyo: JFPRO, 9–27.

Suzuki, S. 1999. "Life and environments in the villages in mangrove areas of Vietnam" (in Japanese). Osaka: Folklore Institute of Kinki University, *KINKIBUNKA*, No. 11, 256–302.

Tsuruda, K. and T. Asano. 1994. "Mangrove reforestation in Vietnam" (in Japanese). In: M. Kogo (ed.). *Report of Basic Research for Cooperating Mangrove Reforestation in Vietnam*. Tokyo: JFPRO, 92–104.

UNESCO. 1981. *Bibliography on Mangrove Research 1600–1976*. Paris: UNESCO.

Photo 15.1 *Kandelia candel* reforestation in Tinh Gio: the seedlings are 6 months old

Photo 15.2 Shrimp farming combined with the planting of *Rhizophora apiculata* at Ben Tre in the Mekong delta

Photo 15.3 Five lodges have been established in Can Gio, Ho Chi Minh City, for education and research regarding mangroves

16
Mangrove forestry research in Bangladesh

A.F.M. Akhtaruzzaman

Introduction

Bangladesh is a small country with an area of about 148,000 km^2. It lies between latitudes 20°34' and 26°28'N and longitudes 88°01' and 92°41'E. Apart from the hilly regions in the north-east and south-east, the country consists of low, flat, and fertile land. Three mighty rivers – the Ganges, the Brahmaputra, and the Meghna – drain a catchment area extending over Bhutan, Nepal, India, Bangladesh, and China. The population of the country is about 125 million and the population growth is 1.7 per cent. The per capita income in Bangladesh is about US$270. The forestry sector generates about 5 per cent of the GDP and the forest area comprises 17 per cent of the land of the country; of this, the Forest Department manages about 10 per cent, the remaining 7 per cent being off-forest land (Anon. 1992); however, only 6 per cent of such land is under closed forest cover. There are three major forest types of natural vegetation in Bangladesh – semi-evergreen forest, deciduous forest, and the mangrove forest.

Mangrove forests

The coastline of Bangladesh is about 710 km long with many tiny islands. Both natural and planted mangroves are available in the coastal area of

Bangladesh (Siddiqi 1999a), which lies between 21–23°N and 89–93°E (fig. 16.1). The natural mangrove forest includes the Sundarbans and the Chokoria Sundarbans. The Sundarbans form the single largest continuous mangrove forest of the world, with an area of about 10,000 km², of which 62 per cent falls within the territory of Bangladesh. The Chokoria Sundarbans has virtually been completely destroyed in recent years as a result of shrimp farming. In addition to the natural forest, a mangrove-plantation programme has been undertaken over an area of 170,000 ha since 1966. *Sonneratia apetala* and *Avicennia officinalis* are the principal planting species for the coastal afforestation (Siddiqi 1999b).

Importance of mangroves

Because of their protective and productive functions, mangroves play a key role in the maintenance of life and productivity of the entire coastal area. Timber, firewood, and other non-timber products are regularly harvested from the mangrove forests. A number of industries (newsprint mills, match factories, hardboard production, boat building, and furniture making) are based on the raw material obtained from the Sundarbans ecosystem. In addition, management of the mangrove forest and mangrove plantations helps generate considerable employment opportunities for the landless, poor, coastal population. However, the intangible benefits from this ecosystem are in no way less important than the tangible benefits. Bangladesh is a cyclone-prone area; the protection offered by the natural mangrove forest to the life and property of the coastal population has led the government to undertake a massive mangrove-afforestation programme all along the shoreline and offshore islands (Siddiqui 2001).

Problems of mangrove forests

Sundarbans

Bangladesh is an overpopulated country. Naturally, there is tremendous pressure on the forest resources, including the Sundarbans ecosystem which supports valuable plant species (table 16.1). Overexploitation of the forest to meet the growing requirements of the people is one of the main problems of the Sundarbans and has resulted in depletion in both the stocking and the productivity of the forest. The growing stocks of *Heritiera fomes* and *Excoecaria agallocha*, the two major species, were depleted by 40 per cent and 45 per cent, respectively, between 1959 and 1983, although the forest is managed under a selection system on a

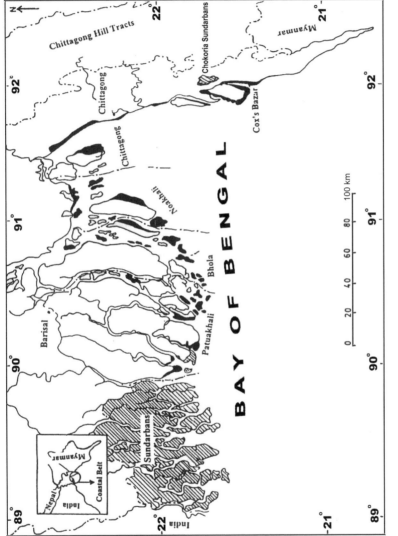

Figure 16.1 Coastal belt of Bangladesh showing location of the planted (■) and natural (▨) mangrove forests. Source: Siddiqi (1999a)

Table 16.1 Economically important plants of the Sundarbans and their uses

Family	Scientific name	Type of plant	Main uses
Avicenniaceae	*Avicennia officinalis*	Tree	Firewood, anchor logs
Combretaceae	*Lumnitzera racemosa*	Small tree	Firewood, posts
Euphorbiaceae	*Excoecaria agallocha*	Tree	Matchsticks and boxes, raw material for newsprint
Leguminosae	*Cynometra ramiflora*	Small tree	Firewood, charcoal
Malvaceae	*Xylocarpus granatum*	Small tree	Furniture
	X. mekongensis	Tree	Furniture, bridges, house construction
Palmae	*Nypa fruticans*	Recumbent palm with underground stem	Thatching for houses
	Phoenix paludosa	Thorny palm	Posts and rafters for huts
Rhizophoraceae	*Bruguiera* spp.	Tree	Furniture, bridge, and house construction
	Ceriops decandra	Shrub or small tree	Firewood, house posts, charcoal
Sonneratiaceae	*Sonneratia apetala*	Tree	Packing cases, panelling
Sterculiaceae	*Heriteria fomes*	Tree	House construction, boat building, electricity pylons and poles, hardboard, firewood

Source: Siddiqi (1998).

sustained-yield basis. The position with regard to stocking has deteriorated further in subsequent years. Other problems of this ecosystem include geomorphological changes, increased water salinity, ecological succession, inadequate regeneration, low yields, and top-dying of *H. fomes*.

Chokoria Sundarbans

The Chokoria Sundarbans initially covered an area of 18,200 ha; previously, this was a moderately dense mangrove forest. Tremendous human pressure, such as overexploitation of forest resources, fishing, cattle grazing, salt production, and other factors, has led to deterioration of the vegetation. Finally, unregulated shrimp farming has practically destroyed the entire mangrove ecosystem. Reforestation in this area with mangrove plants, although now difficult, is still silviculturally feasible with the correct design and planning of shrimp farming.

Mangrove plantations

Bangladesh is a pioneer country with regard to mangrove plantations. The world's largest planting programme is in progress in the coastal areas and offshore islands of the country. However, problems with mangrove afforestation are many and multidimensional. Planting is carried out on newly accreted lands formed from sedimentation of the rivers; the coastal environment is highly dynamic; geomorphological changes are very rapid. Sustainable management of plantations, insect infestation, and lack of natural regeneration are just some of the major problems of the mangrove plantations.

Need for research

Mangroves comprise almost 50 per cent of the forest land of Bangladesh, and the ecosystems encounter very many problems threatening their productivity – and even continuity – in some instances; thus, special emphasis needs to be placed on sustainable management of mangrove ecosystems, including improving the nursery and planting technology. With this end in view, two research divisions – namely, the Mangrove Silviculture Division and the Plantation Trial Unit – have been established under the Bangladesh Forest Research Institute (BFRI). The current research studies under these two divisions are listed below.

Mangrove Silviculture Division studies

- Reproductive biology, and monitoring of the regeneration patterns of the major mangrove species in the Sundarbans.

- Artificial regeneration of main land species in the Non-Commercial Cover (NCC) areas of the Sundarbans.
- Development of nursery and plantation techniques and field performance of *Nypa fruticans* in the natural mangroves.
- Effects of wildlife on the abundance, distribution, and growth of the natural regeneration of major mangrove species.
- Investigation of aided natural regeneration in the Sundarbans.
- Effect of felling and harvesting on the regeneration of *Heritiera fomes* and *Excoecaria agallocha*.
- Selection, propagation, and improvement of the top-dying-resistant *H. fomes* trees in the Sundarbans.
- Diversities and centralization of mangrove species through establishment of an arboretum in the Sundarbans.
- Effect of soil erosion and deposition of silt on the regeneration patterns of *H. fomes* and *E. agallocha* in three salinity zones of the Sundarbans.
- Socio-economic impact of *N. fruticans* extraction in the Sundarbans.
- Nutrient replenishment of *H. fomes* and *E. agallocha* via litter fall in three salinity zones of the Sundarbans.
- Growth performance of *N. fruticans* as affected by various extraction methods.
- Effect of stump height on the coppicing ability of *H. fomes* and *E. agallocha*.
- Investigation of the causes of top-dying of *H. fomes* in the Sundarbans.
- Investigation of the causes of decline in the stocking of *Xylocarpus granatum*.

Plantation Trial Unit studies

- Standardization of planting technique for *Avicennia officinalis*.
- Effect of various thinning intensities on regeneration and succession in *Sonneratia apetala* plantations.
- Verification of site suitability of seven preliminary selected indigenous mesophytic species for different raised coastal areas.
- Study of the performance of *N. fruticans* in newly accreted sites on the central and eastern coastal belt.
- Selection of "Plus Trees" of *X. granatum*, *H. fomes*, and *E. agallocha*.
- Effect of animal browsing, grazing, and trampling on *S. apetala* regeneration.
- Supply of improved planting materials from established seed stands of *S. apetala* species.
- Development of vegetative-propagation techniques for important mangrove species.
- Maintenance and monitoring of previously raised experimental plots with different mangrove and non-mangrove species.

- Study of the performance of existing embankment plantations raised by the Forest Department.
- Development of nursery and planting techniques of some common palms for embankment plantations.
- Study of suitability of different species of palms and different provenance of *Schumannianthus dichotoma* for planting on mounds and dykes at foreshore areas.
- Studies of the effect of social forestry intervention for the improvement of coastal homestead productivity.

In addition to these two divisions, scientists from other disciplines (Soil Science, Forest Botany, Forest Protection, Forest Inventory, Forest Utilization, and Wildlife) are also involved in relevant studies in the natural and man-made mangrove forests of Bangladesh.

Mandate of BFRI

By policy, BFRI has an obligation to provide research support to the Forest Department, Forest Industries Corporation, and others in performing forestry activities. BFRI aims to maintain sustainable productivity of forest land and forest industries without resource depletion. The broad goal is to reduce the demand/supply gap of forest resources. Broad programme areas of research include forestry and nursery management, watershed management, environment and ecosystem conservation, proper utilization of forest products, and others. Priorities are demand-driven research, such as the promotion of farm forestry and agroforestry and the sustainable management of the Sundarbans and other coastal forest land.

Mangrove research achievements

BFRI is an organization of long standing, established in 1955; however, mangrove research was practically initiated in 1985. During the past 15 years, considerable research-based information has been generated: there are as many as 100 research publications on various aspects of mangrove studies and the application of research results to management. These are useful for conservation, raising plantations, and sustainable management of mangrove ecosystems, which are gaining increasing importance. However, it may be noted that the mangrove is a highly dynamic ecosystem: new problems are constantly added to the management of this ecosystem. Thus, research studies aimed at both short-term and long-term problems will be continued.

Notable research findings

The following research work has been conducted:
- Soil characteristics of mangrove environment studied;
- Floral composition of natural mangrove determined;
- Regeneration status in relation to abiotic and biotic factors of the Sundarbans determined;
- Studies on the scope of enrichment planting conducted;
- Vegetation of the Chokoria Sundarbans studied;
- Seed biology of *H. fomes* and *S. apelata* studied;
- Growth performance of ten mangrove species studied;
- Nursery manual for commercially important mangrove species prepared;
- Planting techniques of *S. apelata* and *A. officinalis* developed;
- Nursery, plantation, and management studies for *N. fruticans* conducted;
- Financial rotation for *S. apelata* determined;
- Method for developing second-rotation crops in coastal areas refined;
- Volume tables for plantation species prepared;
- Pests and diseases and their management on mangrove species surveyed;
- Properties and utilization of various mangrove species studied.

REFERENCES

Anon. 1992. *Forestry Master Plan. Asian Development Bank (UNDP/FAO/BGD 88/025)*. Dhaka: Government of Bangladesh. Ministry of Environment and Forests.

Siddiqi, N.A. 1998. "Enrichment planting in the mangrove of Sundarbans – a review." *Bangladesh Journal of Forest Science* 27(2): 103–113.

Siddiqi, N.A. 1999a. "Raising plantations of nipa palm on new accretions along the coastline of Bangladesh." *Journal of Non-Timber Forest Products* 6(1/2): 52–56.

Siddiqi, N.A. 1999b. "Status and conservation of mangroves in the Indian subcontinent." Unpublished paper presented in the *Symposium on Management of Mangrove Ecosystems* held in Bali, Indonesia (7 Sep. 1999), organised by the International Society for Mangrove Ecosystems and Indonesian Institute of Sciences. 13 pp.

Siddiqi, N.A. 2001. *Mangrove Forestry in Bangladesh*. Pr. Chittagong: Institute of Forestry and Environmental Sciences, University of Chittagong, and Nibeden Press Ltd, 201 pp.

17

Socio-economic study of the utilization of mangrove forests in South-East Asia

Kazuhiro Ajiki

Introduction

In recent years, many kinds of impact caused by human activities have severely affected the mangrove areas. Consequently, the total mangrove forest area has been reduced, with resultant deterioration in the coastal environment.

Most studies dealing with mangroves have been conducted in the fields of biology and ecology (Chapman 1976), whereas there are not many studies of the utilization of mangroves and of the practical lifestyles of the local inhabitants. Considering mangroves from a socio-economic aspect to clarify how and by whom mangrove forests have been used, who is concerned with their exploitation, and who has made gains, is now an important issue. Case studies on the micro-scale and field surveys are needed.

The purpose of this study is to describe and compare several types of utilization of the mangrove forests and the effects on the lives of local people in South-East Asia. Special attention is given here to utilization by local villages and to the economic structure of the coastal communities. Field studies have focused on areas in the Philippines, Viet Nam, and Thailand.

Case study in the Philippines (1992)

Study area

The study area is Barangay (settlement) Lincod of the Municipality of Maribojoc, in the south-western part of Bohol Island. According to the census taken in 1990, the total population in Barangay Lincod was 981 and the number of households was 214. The field survey was carried out in August 1992.

In this area, mangrove forests, including nipa (*Nypa fruticans*), formerly covered about 490 ha in the estuary of the Abatan River and nearby coastal zone. Since Barangay Lincod is situated very near to the sea, it has a historical relationship with mangrove forests.

The utilization of mangrove forest by local people

Twenty-six households were selected randomly for questioning (table 17.1). All sampled households are categorized into six types, according to (a) whether they hold farmland or nipa stands and (b) their main source of income.

Seventeen farming households were observed (types I and III in table 17.1). The area of managed farmland is, in general, small: most farmers cultivate about 50–100 ha. Almost all crops are planted only for family consumption; agriculture, therefore, is mainly for subsistence.

As for fishery activities, 6 households catch fish at sea, and 11 families fish in mangrove waters; mangrove areas are, therefore, recognized as important for fishery by local residents.

The utilization of the mangrove forest was analysed. Cutting and using the mangrove tree resources is not sufficient: only three households cut mangrove trees for firewood and only one family makes fish-pens from mangrove wood. Most people said that firewood and construction materials are provided entirely by timber from coconut and other trees.

In this area the utilization of nipa stands is very important. The whole mangrove area including nipa is public land owned by the government. However, customary rights, especially the use of nipa (which is extensively used by local people), have been granted by the DENR (Department of Environment and Natural Resources). Holders of these rights can harvest nipa leaves to make shingles, which are composed of bamboo canes 1.2 m in length and of nipa leaves, and are widely used as materials for roofs and partitions of houses throughout the Philippines.

Of the sampled households, 10 families hold nipa rights (types I and II in table 17.1). The size of holdings varies from 25 to 150 ha. The families cut nipa leaves twice a year, which is a non-destructive, sustainable use.

Table 17.1 Occupational composition in Barangay Lincod, 1992

Type of household	No.	Managed farmland (a)[a] Owned	Managed farmland (a)[a] Tenant	Holding nipa forest (a)[a]	Head of household	Wife	Occupations[b] Children (Male)	Occupations[b] Children (Female)	Others (Male)	Others (Female)	Main source of income A	F	N·NL	FP	W
I. Nipa-holding farming	1	25		150	(56) A·N	(56) H		(24) H·N					✓		
	2	250		100	(50) W·A	(38) H·N							✓		
	3	50		100		(73) N·A							✓		
	4	50		50		(84) H					✓				
	5		200	50	(50) A	(46) H·N	(26) A·N	(48) H·O					✓		
	6	100		25	(65) A	(59) H·N					✓				
	7	25		25	(76) A	(67) H·N		(22) N			✓		✓		
	8	100	300	?	(68) A	(69) O		(30) H	(32) W		✓				
	9	50		?	(70) A	(68) H		(47) H·N	(45) O		✓				
II. Nipa-holding	10			25	(62) A	(70) N					✓		✓		
	11	100				(57) H·O					✓				
	12	100			(42) F·A							✓			
III. Farming	13	100			(62) A	(53) H·O	(30) F·P				✓	✓			
	14	50			(48) A·F	(71) H					✓	✓			
	15	50			(50) W	(47) H·O		(35) W	(37) F		✓				
	16	50			(54) F	(53) A	(27) FP					✓			
	17	50			(50) W	(46) O	(23) W			(21) W					
	18	25													
IV. Nipa labour	19				(47) NL	(40) NL	(16) FP	(15) NL		(70) NL			✓		
	20					(41) NL							✓		
V. Fish-pond labour	21				(45) FP	(38) H	(36·32·22) FP							✓	
	22				(69) FP	(54) H								✓	
	23				(21) FP	(24) H	(24) FP							✓	
	24				(47) FP	(50) FP				(20) H				✓	
VI. Other	25				(28) W	(28) H									✓
	26				(33) W	(33) H									✓

Source: Data by field survey, 1992; from Ajiki (1994).
a. a, are (100 m^2).
b. ages (years) in parentheses; A, agriculture; F, fishery; N, nipa shingle making; NL, nipa labour; FP, fish-pond labour; O, own another business; W, other waged labour; H, housekeeping.

Bundles of cut leaves are carried to each house, where nipa shingles are made by hand. According to the survey, the gross income from selling nipa shingles is 20,000 pesos annually, out of 1 ha of nipa land, which equals the annual food expense of one household. Thus, making nipa shingles occupies a very important position in the life of the people. This is true not only for nipa-holding families but also for other households without nipa rights, since they are employed in the nipa industry (table 17.1).

Effects of the construction of fish-ponds

In the Philippines, zoning and classification of forest land have been repeatedly carried out by the government (Philippine Forest Research Society 1978). As a result of a recent classification based on the Presidential Order 705 of 1978, in Maribojoc, 177 ha of mangroves including nipa stands were designated as "Timberland (permanent forest)," and the other 144 ha as "Alienable or Disposable" land. In the latter case, conversion to fish-ponds was promoted and 25-year lease contracts were made between the Department of Agriculture and the applicants. In Maribojoc, especially near Barangay Lincod, an area of 64 ha has been converted to fish-ponds.

The example of Mr J., who has 50 ha of fish-ponds near Barangay Lincod, was analysed. Mr J. came from Panay Island, where he had had experience in fish-pond management. According to an interview, about 30 per cent by volume of the total production is milkfish (*bangus* in Tagalog) and the other 70 per cent is prawns. As to the employment of labour, the number of contract workers and part-time labour reaches about 250. These workers live in the neighbourhood and are engaged in simple tasks. Moreover, about 35 skilled workers, permanent and professional labourers, are employed in the fish-ponds.

As to the sampled households, there are four families engaged in fish-pond labour only (type V in table 17.1). Of these, three families came from another region. However, not only immigrants are fish-pond labourers: three local families also work in fish-ponds. Work in fish-ponds cannot be ignored, since it is a valuable source of income to the people of the area. On the other hand, nipa shingle makers and fishery workers see the construction of fish-ponds as a great menace. They say that many households have lost the rights of nipa standholding owing to the 1978 classification of forest land and the construction of fish-ponds. Furthermore, they fear that there will be more damage if the fish-ponds are expanded.

Finally, how should the construction of fish-ponds be assessed? As judged on the basis of the sampled households, the main source of income of each family is clearly divided into five types (table 17.1) – namely, agriculture, fishery, nipa shingle making, fish-pond labour, and other waged labour. Whether exploitation of mangrove forests is good or bad cannot be judged hurriedly; in this area it has conferred both benefits and damage on local people. Fish-pond labourers, immigrants, and small farmers without nipa-holding rights gain advantages, whereas nipa-rights holders, labourers, and fishery workers suffer disadvantages.

Case study in Viet Nam (1994)

Study area

The Can Gio District in Ho Chi Minh City is famous for its large, successful, mangrove reforestation effort, sponsored by the government after the Viet Nam War (Nam and My 1992). I tried to relate the current economic condition of the local people to the mangrove-conservation project in Can Gio District.

The field survey was carried out in August 1994. Tam Thon Hiep Commune was chosen for the survey because of its great dependence on mangrove forests. The total number of households in the commune was about 630. In the survey, 21 households were chosen at random for interview (table 17.2): of these households, 12 were in the central settlement and 9 others were in the forest area in newly built houses.

Local people's lives and the utilization of mangroves

First, the economic condition of 12 households in the central settlement is described. These households can be categorized into three groups – namely, poor, middle, and rich class (table 17.2). Four poor families live by crab- and snail-catching only; this is the traditional work in mangrove areas. However, their family income is limited to about 500,000–1,000,000 dongs per month (10,000 dongs $\approx$ 1 US$). Families of the middle and rich classes have other kinds of jobs, such as fishing in the river with nets or traps, shrimp aquaculture, small trading, and brokerage of crabs and snails. The monthly income of five middle-class families is about 1,000,000–2,000,000 dongs per month each. Three rich families earn approximately more than 2,000,000 dongs per month. In general, local people have been dependent on the resources from the mangroves.

Second, I studied six families who have contracts as mangrove protectors with the Protective Forest Management Board (table 17.2). These families have, at maximum, five sources of income (i.e. mangrove planting, thinning, salary for forest protection, aquaculture, and fishing). As a

Table 17.2 Household composition[a] and occupations[b] in Tam Thon Hiep Commune, 1994

Type	No.	Number (1)[a]	Head of household[b]	Wife[b]	Children Male	Children Female	Children Others
Rich	4	7	(45) B	(41) H	(16, 14) B, (5) –	(24, 22) *, (18) B, (12) –	
	7	8	(36) F1	(35) S		(17) S, (16, 14, 12, 10, 4) –	
	8	2	(68) –	(57) S + H		(40, 38, 36, 33, 26) *	
Middle	3	6	(46) F2	(42) F3	(16) –	(23) *, (20) S, (15) F3, (10) –	
	6	5	(34) F2	(32) S	(8) –	(10, 6) –	
	9	9	(47) A	(40) A + H	(19) O, (15, 13) F2 + A, (9) –	(21) –, (17) H, (11) –	
	18	5	(67) A + F1	(66) S	(38, 34, 31, 28) *, (26) A, (20) –	(23) S	
	20	10	(43) F1	(38) H	(17) F1, (15, 11, 8, 6) –	(13, 4, 1) –	
Poor	1	3	(26) F2	(20) F3	(3) –		
	2	2		(60) F2 · 3	(39) F2 · 3		
	5	5		(47) F3	(24) *, (15) F2, (12, 10) –		(90) –, female
	19	9	(36) F2	(34) H	(15) F2, (13, 11, 8, 6, 4, 3) –		

Protection contracted	10	4+5	(46) FM+A+F1	(45) H	(23, 22, 17) FM+A+F1, (14) –	(16) H, (12, 9) –
	11	7	(60) FM+F1·2	(52) H	(36, 32, 28) *, (26, 15) FM+F1·2, (20) O	(30, 22) *, (18) FM+F1	(18) H, female
	12	7+3	(40) FM+F1·2	(39) H	(22, 20) FM+F1·2, (18) F1·2, (16, 12, 7) –	(14) H	(18) H, female
	13	8	(57) FM+A+F1·2	(52) A+H	(28, 26) *, (16, 14) FM	(35, 30, 22, 21) *, (19) FM, (18) H, (10, 9) –
	14	7	(50) FM+A+F2	(44) FM+A	(18) FM+A+F2	(9, 7, 4, 3) –
	21	6?	(?) FM+A	(39) FM+A	?	?
Shrimp culture	15	7	(57) A	(54) H	(36) *, (25) A+F2	(32, 30, 23) *, (21) A, (19, 17) H, (11) –
	16	2+3	(31) A	(28) H	(12) *	(4, 1) –
	17	2	(41) A	(41) A		(22, 20, 18, 16, 14) *	(22) A, male

Source: Field survey, 1994.

a. Number (1) means number of family members living together; if they have two houses, total number is shown using '+'. Ages shown in parentheses.

b. Occupations are shown as follows: F1, fishing on river; F2, crab catching; F3, snail catching; A, aquaculture; FM, forestry; S, store management; B, broker; O, others; H, housekeeping; –, student or no job; *, moved away.

result, their monthly income is 1,000,000–2,800,000 dongs, although the differences between the families are fairly large. Thus, protection contracts make it possible for these families to live to the standard of the middle class of the local area or above.

Interviews were also carried out with three other families living in the forest with shrimp-ponds. This type of aquaculture management is non-intensive. These families can earn an income almost equivalent to that of the rich families.

The forest-protection contract system with land allocation which came into force after 1990 contributed effectively not only to the conservation and production of mangrove forests but also to poor people's welfare. However, there still remains the problem of managing large mangrove forest areas and of strictly preventing illegal felling.

Case study in Thailand (1996–1997)

Study area

The study area was in the southern part of Thailand on the Andaman Sea coast at Amphoe (District) Khlong Thom in Changwa (Province) Krabi. The first field survey took place from November to December, 1996 and the second one from July to August, 1997. I mainly used interview surveys with mangrove charcoal-burners and residents in the sampled settlement.

Mangrove forest-management policy

In Thailand, all mangrove forests are national forests and are managed by the Royal Forest Department (RFD). The management policy can be summarized as follows.

In 1968, the RFD revised its mangrove-management plans. The silvicultural system applied was clear felling. The rotation was set at 30 years, with a felling cycle of 15 years, and is practised by dividing the area into 15 coupes (cutting areas), each of which is further divided into strips 40 m wide. Alternate strips are cut every 15 years, thus giving a rotation of 30 years. The period of a concession is fixed by the issue of a long-term permit, lasting 15 years. The concessionaires have to protect the forest under their concession and improve the exploited areas by reforestation at their own expense (Aksornkoae 1987; Havanond 1995).

This silvicultural system has, for a long time, enabled the concessionaires to produce mangrove charcoal. In August 1996, however, the Cabinet made a new resolution that prohibits all types of mangrove felling throughout Thailand. This decision has been in force since December 1996.

Human impact on mangroves by local people

For intensive survey, the Ban (Settlement) Tha Pradu of Tum Bon (Village) Huai Nam Khao in the Amphoe (District) Khlong Thom was chosen. Of the total 105 households in the settlement, 20 were randomly selected for interview, with the results summarized in table 17.3.

These people are mainly engaged in agriculture and fishing: 13 families hold farmland planted with rubber or oil palm, and 15 families have boats mainly used for fishing; 2 families (nos. 2 and 20 in table 17.3) have managed charcoal factories.

Only two families said that they rarely use mangrove wood. When mangrove wood is needed, permission must be sought from the mangrove concessionaire. The families use *Aegialitis rotundifolia* for firewood; *Avicennia alba* and *A. officinalis* as building poles; and, occasionally, *Bruguiera parviflora*, *B. cylindrica*, and *Aegialitis rotundifolia* as poles for livestock pens. There are three reasons why the use of mangrove wood for fuel is decreasing. First, every mangrove is protected by the concessionaires, who permit others to cut mangroves for wood only in times of extreme need. Second, cooking-fuel for domestic use has been rapidly changing from charcoal or wood to propane gas. Table 17.3 shows that, although charcoal is also important for supplementary uses, 15 of the total number of families use propane gas for cooking: in this settlement a so-called fuel revolution is under way. Third, mangroves are recognized as an important resource for fishing, in which many people are engaged. Thus, it is concluded that any human impact on mangroves by each local family can be disregarded at present. However, as it is known that mangroves have an important role as nurseries for many kinds of fish, many locals fear that water pollution in mangroves caused by shrimp-ponds that operate behind the mangrove areas may damage open-water fisheries in the long term.

Conclusions

Comparing the results of the surveys at three selected sites in three different countries, it is clear that the government policy on the management of mangroves has the greatest impact and its effects are most important in all cases. However, the lifestyles and habits of local people differ in type and degree of dependence and utilization of the mangroves from place to place. As well as people from different places being at

Table 17.3 Household composition and occupations in Ban Tha Pradu, 1997

Type	No.	Number (1)[a]	Occupations[b] Household	Wife	Male	Female	Farming land (ha)	Crops	Boat	Main source of income	Fuel for cooking[c] Firewood	Charcoal	Gas
I	1	7	(38) A+F	(27) A+F	(5)(3) –		16	Oil palm, rubber		A+F		◎	
	2	4	(41) A+F	(33) F+H			16	Oil palm, rubber	5	A+F			◎
	3	7	(47) A+F	(38) T	(22) A+F, (17) –	(19) T+H, (10) –	4.3	Oil palm	1	A+F+T	△	○	◎
	4	4	(35) A+F	(34) H	(5) –	(11) –	3.2	Oil palm	1	A+F			◎
	5	8	(50) A+F	(45) A+F	(21)(20)(17) A+F (14) (9) –	(11) –	2.2	Rubber, oil palm	4	F+A		○	◎
	6	4	(30) W+F +A	(28) H	(3) –	(7) –	1.8	Rubber, oil palm	1	W+F		○	◎
	7	8	(53) F+A	(45) H	(26)(23) F+A (12) –	(26)(20)(17) H	1.5	Rubber	3	F+A	△	○	◎
	8	2	(60) W+F +A	(60) –			0.8	Rubber		W+F	●		
	9	7	(52) F	(43) W	(14) –	(22)(21)(19) H (12) –	0.2	Rubber	2	F+W		○	◎
II	10	6	(60) A	(49) H	(25) A, (12) –	(17) H, (10) –	3.5	Rice, palm, rubber	1	A	●	△	○
	11	6	(45) A+T	(32) H	(15)(12)(10) (0) –	(13) –	1.1	Rubber		A+T	△	○	◎
	12	5	(39) A	(33) T+H	(10)(1) –	(7)(4) –	0.6	Oil palm		A+T		○	◎
	13	7	(40) A+W	(37) A	(17)(11)(10) –		0.5	Rubber		A+W	●	○	◎

266

III	14	9	(55) F	(49) F	(31) F, (13) –	(29) (20) (16) H (7) –	2	F	◉
	15	3	(42) F	(42) H	(9) –	(22) W	1	F	
	16	6	(65) F	(59) F	(25) W	(7) –	1	W+F	◉
	17	4	(32) W+F	(32) H	(2) –		1	W+F	◉ △
	18	5	(68) F	(61) F	(21) F	(17) F	1	F	◉ △
	19	5	(53) F	(49) F		(21) F, (14) –	1	F	◉ ○
IV	20	2	(48) C+T	(48) C+T			12	C+T	◉

Source: Ajiki (1999); data source: field survey 1997.
Type I: farming and fishing households, II: farming households, III: fishing households, IV: others.
a. Number (1): number of family members living together.
b. Ages shown in parentheses. Occupations A, agriculture; F, fishing; C, charcoal making; T, trade; W, waged labour; H, housework; –, student or no job; children (female) include child's wife.
c. Fuel for cooking: ◉ main use, ○ second use, △ third use.

various levels of dependence on the mangrove forest, they also receive different types of government assistance.
1. In the Philippines, capitalists have exploited mangrove forests and converted them to fish-ponds, an activity promoted by the government. On the other hand, local people have sustainably used mangrove resources. Hence, a conflict over the utilization of mangroves was observed in the study area.
2. In the case of the Can Gio District in Viet Nam, the government has strongly supported the plans for reforestation and conservation of mangroves. Some local people can achieve bonuses when they participate in the planning. Moreover, many families depend on the planted mangroves, since they have traditionally utilized the mangroves in a sustainable manner.
3. In Thailand, the government in the past gave concessions to a sector of people to use mangrove forests for charcoal production. As a result, in the study area, mangrove forests have been maintained, and play an important role for local consumption and commercial fishing by local people.

In socio-economic studies dealing with the utilization of mangroves, the government policy should first be well defined – as should be the uses to which the mangroves have been put in the past, are put in the present, and will be put in the future. It is strongly recommended that the focus should be on the local people's way of life and life standards. Micro-scale studies and research based on field surveys is the key to the success of sustainable management of mangrove ecosystems.

Note

This chapter was compiled from detailed published papers by Ajiki (1994, 1996, 1999), with revisions and additions.

REFERENCES

Ajiki, K. 1994. "The decrease of mangrove forests and its effects on local people's lives in the Philippines." *ISME Mangrove Ecosystems Proceedings* 3: 43–48.
Ajiki, K. 1996. "Economic condition of local people and mangrove conservation in Can Gio, Ho Chi Minh City." In: Phan Nguyen Hong and Phan Nguyet Anh (eds). *Proceedings of the National Workshop: Reforestation and Management of Mangrove Ecosystems in Vietnam: Do Son, Hai Phong 8–10 October, 1995*, Hanoi: MERC/ACTMANG, 118–120.
Ajiki, K. 1999. "The human impact on mangrove forests in Southern Thailand –

Findings from field survey in Changwat Satun and Krabi." *TROPICS* 8: 233–237.

Aksornkoae, S. 1987. "Mangrove Resources." In: Anat Arbhabhiramam, Dhira Phantumvanit, John Elkington, and Phaitoon Ingkasuwan (eds). *Thailand's Natural Resources Profile: Is the Resource Base for Thailand's Development Sustainable?* Bangkok: Thailand Development Research Institute, 145–163.

Chapman, V.J. 1976. *Mangrove Vegetation*. Leutershausen (Germany): Strauss and Cramer.

Havanond, S. 1995. "Mangrove resource management in Thailand." In: Choob Khenmnark (ed.). *Ecology and Management of Mangrove Restoration and Re-generation in East and Southeast Asia. Proceedings of the ECOTONE, 18–22 January 1995, Wang Tai Hotel, Surat Thani, Thailand*, 177–183.

Nam, V.N. and T.V. My. 1992. *Mangrove for Production and Protection – A Changing Resource System: Case Study in Can Gio District, Southern Vietnam*. Bangkok: FAO of the UN.

Philippine Forest Research Society (ed.). 1978. *Forum on Grassland and Mangroves: Conservation and Utilization Issues*. Laguna (Philippines): Philippine Forest Research Society.

18

Sustainable mangrove management in Indonesia: Case study on mangrove planting and aquaculture

Atsuo Ida

Introduction

The project

The proposal for a mangrove-rehabilitation project was submitted by the Government of Indonesia to the Government of Japan in 1990. In response to this proposal, the Japan International Cooperation Agency (JICA) dispatched four survey teams to Indonesia. Subsequently, JICA and the Directorate-General of Reforestation and Land Rehabilitation (DGRLR, currently known as the Directorate-General of Reforestation and Land Rehabilitation and Social Forestry) of the Ministry of Forestry (currently known as the Ministry of Forestry and Estate Crops) of Indonesia agreed upon the implementation of the Development of Sustainable Mangrove Management Project for five years in November 1992.

JICA and DGRLR jointly executed the final evaluation in July 1997 and agreed on the continuation of the project as a follow-up scheme for two years to complete the remaining studies. Accordingly, the project started in December 1992 and was terminated at the end of November 1999.

Objectives

The objectives of the project were to establish a sustainable mangrove management system, aiming not only to conserve mangrove forests but at the same time to maintain a good balance between conservation and economic activities. The project also aimed to develop a silviculture technique of mangrove species in deforested and/or degraded areas, which includes development of technology as well as promotion of mangrove rehabilitation through economic activities. These objectives were achieved through the collection of basic, relevant data; data analysis; and the development and establishment of fundamental technology.

The ultimate target of the project was to explore the feasibility of sustainable mangrove management.

Implementation

In order to achieve these objectives, JICA dispatched six experts (a team leader, a liaison officer, and experts in nursery, silviculture, ecology, and forest management) in the first period, and four experts (a team leader, a liaison officer, and experts in silviculture and forest management) in the follow-up period.

The activities of the project were as follows:
- Selection of tree species for mangrove plantations;
- Development of silvicultural techniques;
- Estimation of the cost of mangrove plantations;
- Estimation of the effects of mangrove forests on the surrounding environment;
- Conservation management and assessment of flora and fauna in the mangrove ecosystem at the project site;
- Pest and disease control techniques;
- Study of the social and economic benefits from forestry and fisheries in mangrove forests and surrounding areas;
- Preparation of a mangrove forest management model;
- Development of utilization techniques for mangrove forest products.

These activities were divided into 26 elements for experiments and studies which were executed by JICA experts and their Indonesian counterparts. The results are reflected in the publications *Nursery Manual*, *Silviculture Manual*, *Handbook of Mangroves*, and *Sustainable Management Models*, all of which were published by the project.

Products (fig. 18.1)

The major products of each component are as follows:
- Nursery: *Nursery Manual*, and creation of a permanent nursery;
- Silviculture: *Silviculture Manual*, and raising a 254 ha plantation;
- Ecology: *Handbook of Mangroves*, and a herbarium collection of species;

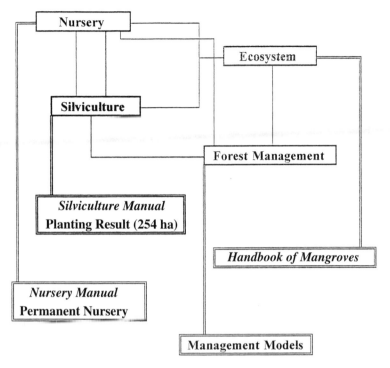

Figure 18.1 Relationships between, and major products of, each component of the project

- Forest Management: *Sustainable Management Models*;
- Volume and yield prediction tables for *Rhizophora apiculata*.

Summary of published outcomes

Nursery manual of mangrove species at Benoa Port in Bali (fig. 18.2)

The *Nursery Manual* is based on the results of practical nursery procedures and studies made during the course of the project. Its aim is to give field technicians the required knowledge, orientation, and guidance.

The selection of species was based on such factors as their distribution in natural forests and seed availability. The manual describes the establishment of nurseries, nursery practices, working schedules, out-planting, and other details. Should the manual be used in other areas, it will be essential to investigate important local conditions, such as salinity and

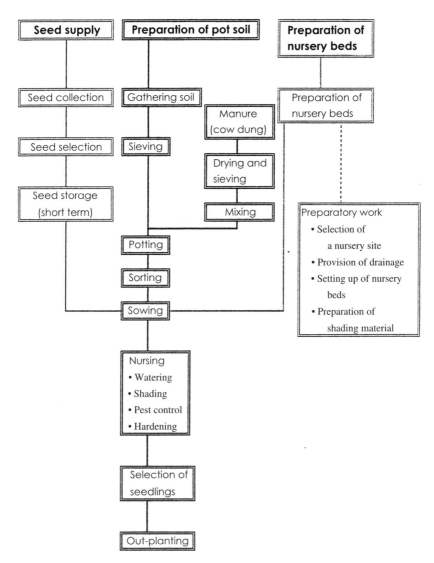

Figure 18.2 Flow chart of nursery work: *Nursery Manual* for mangrove species at Benoa Port, Bali

type of soil, and the methods outlined should be modified according to local conditions.

The following seven species were selected:
- *Rhizophora mucronata*: viviparous seeds;
- *Rhizophora apiculata*: viviparous seeds;

- *Bruguiera gymnorrhiza*: viviparous seeds;
- *Sonneratia alba*: normal seeds;
- *Avicennia marina*: cryptoviviparous seeds;
- *Ceriops tagal*: viviparous seeds;
- *Xylocarpus granatum*: normal seeds.

Silviculture manual for mangroves in Bali and Lombok (fig. 18.3)

The manual was formulated on the basis of the results and experiences of the project. It gives foresters and field technicians the required knowledge and guidance for silvicultural activities in abandoned shrimp-ponds or logged-over former coral seashores.

The selection of species was based on their distribution in natural forests, seed availability and so on. The manual describes practical working procedures, beginning with a preliminary investigation up to the end activities. In the event that the manual will be used in other areas, it will be essential to investigate a number of important local conditions, such as ground height level and the phenology of mangrove species; the manual should be modified according to local conditions.

The eight species selected were as follows:
- *Rhizophora mucronata*; Bali and Lombok;
- *Rhizophora stylosa*; Lombok;
- *Rhizophora apiculata*; Bali and Lombok;
- *Bruguiera gymnorrhiza*; Bali and Lombok;
- *Sonneratia alba*; Bali;
- *Avicennia marina*; Bali;
- *Ceriops tagal*; Bali;
- *Xylocarpus granatum*; Bali.

Handbook of mangroves in Indonesia (Bali and Lombok)

The handbook covers 30 true mangroves (18 major and 12 minor components) and 20 mangrove associates (coastal plants) that are commonly found in Bali and Lombok. Most of these species are also identified in other parts of Indonesia and in other Asian or Pacific Ocean countries where the environment is suitable for mangroves.

The handbook is designed for all those who are interested in mangroves. It is a practical guide that does not require any special botanical knowledge of plant classification because there are many photographs, illustrations, key matrices with graphical icons, graphical habitat indicators, local name lists, and simple descriptions. All aids in the handbook would help users distinguish species more easily in the field.

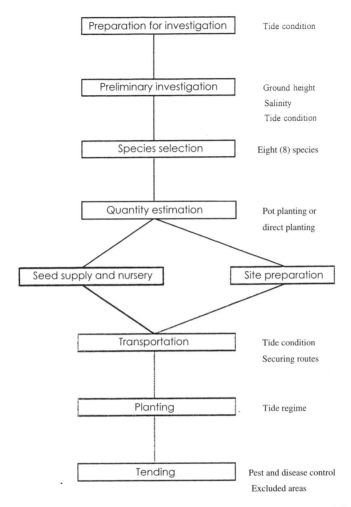

Figure 18.3 Flow chart of silvicultural activities: *Silviculture Manual* for mangroves in Bali and Lombok

Sustainable-management models for mangrove forests

For the preparation of sustainable mangrove forest-management models, the project focused on the following objectives:
- To examine the feasibility of investments capable of undertaking reforestation alongside profitable economic activities;
- To establish sustainable systems for mangrove forest management adapted to both local conditions and management purposes.

Table 18.1 Examples of mangrove planting and traditional extensive aquaculture[a]

	Silviculture system	
	Sidoarjo	Curah Sawo
Planting species	*Avicennia marina*	*Rhizophora mucronata*
Planting sites	Banks, shoreline, and river-banks	Banks, shoreline, and river-banks
Planting method	Pot planting After 20 years, replanting Banks: 5 m intervals Others: 1 m × 1 m	Pot and direct planting Banks: 1 m intervals Others: 1 m × 1 m
Purpose	• Maintenance of banks, shoreline, and river-banks • Used for salinity control and as fodder • Used as firewood • Used for land reclamation and construction of new ponds	• Maintenance of banks, shoreline, and river-banks • Use branches for firewood and poles from thinning

	Procedure	
	Sidoarjo	Curah Sawo
Harvest	Shrimp and milkfish	Shrimp and milkfish
Procedure	1. Dry and clean up ponds in the dry season, repair of water gates 2. Replenish with water and await seaweed and algal growth 3. After full growth, empty ponds and dry seaweeds and algae	1. Dry and clean up ponds in the dry season, repair of water gates 2. Replenish with water

4. Refill pond with water, after new growth of algae and seaweeds plankton will also increase in abundance
5. Stock pond with $10-15 \times 10^3$ fry/ha and 1×10^3 young milkfish/ha
6. Add leaves and branches of *A. marina* or *Acanthus ilicifolius* to the ponds for salinity control in the rainy season
7. Harvest after 3–4 months. Survival rate: shrimp 40–60%; milkfish 80%
8. After harvesting, flood the pond with water and stock with fry and milkfish

3. Stock with 5.5×10^3 fry/ha and 0.5×10^3 fry/ha and 0.5×10^3 young milkfish/ha
4. Harvest after 3–4 months. Survival rate: shrimp 30–40%, milkfish 80%
5. After harvesting, flood the pond with water and stock with fry and young milkfish

Other species

Species: Black tiger shrimp
2–3 times yearly
A. ilicifolius can be added to the water to produce feed for the plankton, if extended feeding becomes necessary.

Species: white or banana shrimp
2–3 times yearly

---

a. Two types of aquaculture systems are practised in Indonesia. For comparison, examples are drawn from two villages – Sidoarjo and Curah Sawo in East Java.

Conservation and preservation of mangrove forests cannot be achieved if implementation neglects the people's desire to seek a better living. The models were formulated for guaranteeing sustainable land use and people's welfare on a long-term basis. Flow charts should be based on field data to ensure sustainable utilization and conservation of mangrove forests. Therefore, the yield prediction and volume tables for the mangrove species (*Rhizophora apiculata*) most utilized in Indonesia were prepared following the analysis of collected data. The analysis revealed that there were at least two different types of mangrove forests based on their growth rate, which are high-growth and low-growth sites. According to the results of social economic surveys, management models can be divided into those for low and high labour-intensity work. Consequently, four sustainable mangrove forest-management models were categorized, as formulated by the project. These four categories were as follows:

- *Model A: Chip Production based on the Selective Cutting and Reforestation System.* Low labour-intensity work in high-growth sites.
- *Model B: Charcoal Production based on the Selective Cutting and Reforestation System.* High labour-intensity work in high-growth sites.
- *Model C: Charcoal Production based on the Clear Cutting in Small Area and Plantation System.* Low labour-intensity work in low-growth sites.
- *Model D: Combination of Extensive Aquaculture and Mangrove Planting.* High labour-intensity work in low-growth sites.

Mangrove planting and aquaculture

The sustainable mangrove-management models were formulated on the basis of existing management cases which were collected through field surveys in Indonesia. One of the most important findings is model D, which combines extensive aquaculture and mangrove planting. This model underlines the concept of simultaneous sustainable aquaculture and sustainable production of the mangrove forests (table 18.1). Sustainable extensive aquaculture has been operated by local people in Indonesia for more than 300 years. The local people understand the importance of mangrove forests and mangroves' functions and are willing to plant mangrove trees for the maintenance of their living conditions and future profits.

This extensive aquaculture is uncultivated and unfed. No hatcheries or additional food need be added to the ponds; on the contrary, the system is operated through the maximum utilization of the existing natural resources of mangrove areas. Some examples of the operations are as follows:

- Utilizing drying seaweed/algae to nourish plankton;
- Feeding shrimp and milkfish at the same time;
- Planting *Avicennia* sp. or *Acanthus ilicifolius* in the ponds to control salinity and pH in the rainy season;
- Planting *A. ilicifolius* in the ponds to extend the feeding period.

Conclusions

The published results of the project, such as the handbook and two manuals, were compiled on the basis of existing knowledge in Indonesia and include the results of studies and experiments achieved during the cooperation period. They are designed to help foresters and field technicians to implement various field surveys and rehabilitation/reforestation activities.

The four models indicate achievable ways to accomplish sustainable mangrove forest-management systems, as exemplified by types of practical management in Indonesia. They are to be modified and categorized for ease of understanding.

Thus, under certain conditions, sustainable mangrove-forest management is shown to be achievable. The results of the project are also expected to be utilized to enhance rehabilitation/reforestation activities and to maintain existing mangrove forests in a sustainable manner. In addition, this recognized extensive aquaculture system and the related techniques should be investigated, and the sustainability should be confirmed, on the basis of experimental scientific studies.

Finally, it is anticipated that both JICA and the Ministry of Forestry and Estate Crops in Indonesia will develop and disseminate the products and publications of the project in the near future for the purpose of conserving mangrove forests and enhancing the lives of the inhabitants throughout the tropical coastal belt.

19

Sustainable use and conservation management of mangroves in Zanzibar, Tanzania

Thabit S. Masoud and Robert G. Wild

Introduction

Zanzibar is a semi-autonomous state within the United Republic of Tanzania. For most sectors, including natural resources and environment, it has separate laws, ministries, and departments to that of mainland Tanzania. Zanzibar, in common with many developing nations, is endowed with rich natural resources supporting an increasingly growing human population. Currently, the country is at a crossroads as it attempts to move beyond traditional land-use patterns and, in the face of the doubling of the population over the next fifty years, wishes to reverse the trend of increasing poverty. Current efforts include broadening the economic base beyond agriculture and natural resources and into other areas (particularly tourism), as well as increased biodiversity conservation and natural-resources management. Some of these efforts are in conflict.

The total population of the islands was 800,000 in 1998, with a density of 260 people per square kilometre. The majority of people depend on the utilization of natural resources, key to which are reef fisheries, forests, and agriculture. The forest estate includes significant areas of mangrove. Although the natural resources are deteriorating all over the island, mangroves have probably fared better than coral reefs and coral-rag forests (Mohammed 1999). While heavily exploited mangrove forests, by and large, have maintained the capacity to recover, the area lost to other land uses is limited, although salt extraction is having an increasing impact. Community participation in management of natural resources is now emphasized. Regarding forestry, significant steps have been made to empower communities for natural-resource management, including that of mangroves. Zanzibar is now poised to make community-based mangrove management the rule rather than the exception. The Commission for Natural Resources has sufficient well-trained staff, but the lack of central government financial support is now the limiting factor to making this a reality.

Zanzibar mangroves

The Zanzibar mangrove forests cover most of the shallow water coasts of Zanzibar, protecting about 1,000 km of the coastline (FAO 1995). Zanzibar – which is made up of two main islands, Unguja and Pemba – has mangroves of comparatively recent origin (Griffith 1950), growing mostly on the mud washed down by erosion or the sand thrown up by the sea. Heavy human impact is considered to have reduced the quality of Unguja mangroves, although climatic and edaphic factors may have made a significant contribution (Shinula 1990). Perhaps for a similar reason, Pemba mangroves grow faster than those on Unguja (Griffith 1950).

Although the two islands were once completely forested, forests now cover only 27 per cent of the land area. Of this remaining forested area (132,437 ha), 74 per cent has developed on the raised coral limestone (coral rag) as coastal evergreen coral-rag forest. The 3 per cent of this area that remains as high forest is legally gazetted as forest reserves, the remaining 71 per cent having been somewhat degraded to coral-rag thicket and bush. The coastal evergreen forest of the coral rag has very high biodiversity values, containing rare and endangered forest and wildlife species of both national and international significance (Pakenham 1984; Krebs 1997). Mangroves are the second largest formation and constitute 15 per cent of this forest cover. Of the forest area, 26 per cent, including all mangroves, is protected as forest reserve (FAO 1995); the rest falls under the public (open access) land mainly in the coral rag (fig. 19.1 and table 19.1).

Zanzibar has some 19,748 ha of mangrove forests, of which 5,829 ha grow in Unguja and 13,919 ha are found in Pemba (Leskinen, Pohjonen, and Ali 1997). Nine species have been reported to occur in Zanzibar: these are *Avicennia marina, Rhizophora mucronata, Bruguiera gymnorrhiza, Ceriops tagal, Sonneratia alba, Xylocarpus granatum, X. moluccensis, Lumnitzera racemosa,* and *Heritiera littoralis. X. moluccensis* occurs in only one locality – Micheweni creeks in Pemba (Shinula 1990).

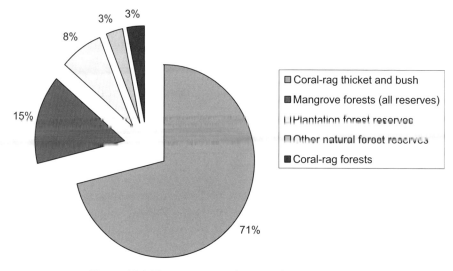

Figure 19.1 Forest types and proportions in Zanzibar

Management history, legal background, and evolution

The mangroves of Zanzibar have historically been exploited for tannin, building poles, and fuel. Bark extraction, for tannin, was an important revenue earner: between 1938 and 1947, a total of 28,786 tonnes of bark was exported to Europe and America (Griffith 1950). Unfortunately, bark was not extracted with sustainable plans in mind, and the mangroves (particularly on Unguja) were seriously damaged (Nasser 1994).

In 1945 the mangrove forests were recognized as government land ac-

Table 19.1 Forest areas of Zanzibar

Areas	Extent (ha)	Total (%)
Natural reserves[a]	3,968	3
Plantation reserves[a]	10,231	8
Coral-rag forest	3,833	3
Coral bush and grassland	94,657	71
Mangroves[a]	19,748	15
Total forest area	132,437	100

Source: FAO (1995); Leskinen, Pohjonen, and Ali (1997).
a. Legally protected.

cording to the wood-cutting decree. Communities had no rights of ownership, but concessions for woodcutting were granted to local people for domestic purposes. In 1959, all the mangrove forests were declared forest reserves under the Forest Reserve Decree, Cap. 120 of 1950; by 1965, all the mangrove forests had been gazetted, and the management fell directly under the Forest Department. The Forest Department, currently known as the Commission for Natural Resources (CNR), exercised control over exploitation mainly by the following methods:
- Alternate closing and opening of areas on a 10-year rotation basis;
- Limiting the harvesting of building poles less than 10 cm in diameter;
- Prohibiting the use of mangrove for firewood and bark removal;
- Issuing cutting and transportation permits for any mangrove extraction;
- Law enforcement.

Over the last 50 years, and in spite of extensive illegal felling, these actions have, to a certain extent, been effective. Although illegal has exceeded legal felling many times over (FAO 1995) and has reduced the current availability of mangrove construction materials to almost zero, Zanzibar's mangrove ecosystems are, however, largely intact, albeit overcut. Some opening and closing systems are still operational; a ban on firewood cutting in mangroves is largely respected; and, in many places, mother trees are numerous, allowing for healthy regeneration (Nasser 1994). The integrity of the mangroves is a reflection of other coastal and marine resources which have been regarded as largely intact and in good condition, compared with those in neighbouring countries (COLE 1996).

Threats to management

Although figures are not available, some losses of mangrove patches have occurred owing to removal to make way for industrial facilities and hotel development or for commercial woodcutting, and also from pollution (by an oil terminal), but these are considered to make up only a small proportion of the total area of mangrove forests. Whereas Zanzibar's mangrove ecosystems have not been overly damaged, in comparison to some other countries, the indications are that serious degradation and land-use changes are imminent. The following trends could precipitate these irreversible changes unless preventative and corrective actions are taken immediately: increased demand from urban areas for wood; non-mangrove forest degradation; declining agricultural yields and fish catches; lack of alternative sources of livelihood; decline in income from traditional crops (clove and coconut); and increased human population.

In response to these trends, the government of Zanzibar has, in recent years, been advocating new income sources to improve the local economy. Salt making (by solar evaporation pans in mangrove areas) has been seen as one of several alternatives, but has led to the clearing of at least 100 ha of mangrove forest. Some of these projects have been financed by UN agencies, as income-generation projects, with belated, ad hoc approval from the regulatory agencies (including the CNR) in the absence of a clear policy for conservation, utilization, and land-use change in mangrove areas.

Conservation initiatives

In response to these ongoing threats to the mangrove system, a number of conservation initiatives have taken place.

Pilot approach to community-based management at Kisakasaka Village

Kisakasaka Village is located some 10 km from Zanzibar Town. With a population of 526 people relying entirely on 400 ha of mangrove for their livelihood (CNR 1997), Kisakasaka was selected by the CNR as a pilot village for local management practices in Zanzibar. The need for community participation was seen as early as 1993; when community members were visited, the majority showed their interest in mangrove management.

Through facilitation by the CNR staff, a village-based Conservation Committee of 10 men and 5 women was formed, responsible for day-to-day management on behalf of the larger community and to prepare a draft management plan and by-laws with facilitation by the CNR staff. Field surveys were carried out under the guidance of the Conservation Committee. The general pattern was overuse of mangroves close to settlement areas. In less accessible areas, there were large trees (mainly *Rhizophora mucronata*) exceeding 30 cm in diameter, with a 15 m canopy.

Field surveys helped the community and the CNR to identify and agree on priority areas for full protection and those that could be put under rotational harvesting. The Kisakasaka mangrove-management plan was simple – to regulate harvests and to act as a memorandum of understanding (i.e. a legal agreement) between the CNR and the community. Community views and recommendations were collected through a series of meetings and workshops within and outside the village. The main features of the plan were as follows:

- Kisakasaka mangroves were divided into two blocks – a use zone and a full protection zone;
- The use zone was to be harvested on a 5-year rotation, based on user experience;
- Although felling of mangroves for firewood is illegal, an exception was made for the Kisakasaka village community, and each individual was allowed to burn one charcoal kiln each month;
- Tree cutters were obliged to pay both cutting and transport permits, to both CNR and the community. Parallel receipts were to be issued by CNR and the Conservation Committee;
- Alternative income-generating activities were to be sought to reduce pressure on the mangrove ecosystem.

Progress to date

The pilot community empowerment on mangrove management at Kisakasaka, although not devoid of problems, can be considered to be successful and reflects the committment of the community when entrusted to managing their own forest areas. Most of the current problems do not really derive from the community – meaning that, with proper support and guidance, the community can perform a far better job than the government.

The few problems experienced during implementation of the plan include the following:
- Inadequate governmental resources to monitor implementation;
- Loss of government revenue (the revenue collector – District Forest Officer (DFO) West – visits the village only once a week);
- Weak collaboration between the three institutions interested in working with the community on Kisakasaka mangrove management. WWF (Menai Bay Project) is currently trying to pull together the efforts of a local NGO (JMZ), as well as that of the government, to agree on a common action plan for the management;
- The finance committee in Kisakasaka is separate from, and not the same as, the conservation committee; much of the time of the finance committee is taken by revenue collection rather than in helping to protect the resource base.

Signs of overharvesting in localized use-zones have been observed. This has been realized by the community itself and is probably due to population increase, as the village has been receiving a large number of immigrants from mainland Tanzania (also in need of mangrove resources). Strategies to overcome the situation include the use of tree stumps and roots for charcoal-burning and simultaneous cutting and replanting. In addition, the Village Conservation Committee stresses review of the mangrove-management plan to be appropriate to the current situation.

Improved policy and legislation

In recognizing the need to bring policy and legislation in line with the new community-based approaches to forest management, the Commission embarked on a process of developing both. The Kisakasaka pilot activity provided practical experience while legislation was being developed. In 1996, both policy and legislation were approved, giving Zanzibar the most progressive forest-management instruments in East Africa.

During the process of formulating the new forestry policy, a separate policy for mangroves was found to be justifiable (FAO 1995). The policy requires mangrove management to be integrated across all relevant sectors (GOZ 1996). In recognition of the fact that nearly 45 per cent of Zanzibar's total population in 1993 lived in 63 coastal villages (COLE 1996), community participation in the management of mangrove resources was re-emphasized. Implementation of the new mangrove policy directed the government to reduce "policing" and to enhance participation.

Following the new policy directives, the new forestry legislation (Forest Resources Conservation and Management Act – FRECMA) was enacted in 1996 and provided for a legal backing of community empowerment over forest-resources management including mangroves. A local management plan on each of the designated community-managed areas would be required for presentation to the CNR to act as a legal agreement between the commission and a respective community group.

Coastal-zone management, conservation, and development at Chwaka Bay

Chwaka Bay is the site of the largest single stand of mangroves on Unguja. It has been the focus of two separate conservation initiatives – namely, a pilot planning site for integrated coastal-zone management, and conservation work through an integrated conservation and development project.

During the 1940s, the Chwaka Bay mangrove was under traditional management by the local council and divided into two large management blocks. The council was formed by the eight villages around the bay. By-laws were established to control harvesting by villagers. Community management continued until 1965, when all the mangrove areas became Forest Reserves.

Integrated Coastal Area Management (ICAM) programme

In recognizing the dependence of coastal communities on mangroves and other marine resources, the Government of Zanzibar (Department of Environment) has developed an Integrated Coastal Area Management (ICAM) pilot programme. This was supported by technical cooperation of the United Nations Environment Programme (UNEP) and FAO and facilitated by the Coastal Resources Center, University of Rhode Island (USA). An action strategy was prepared for the Chwaka Bay–Paje pilot site, and mangroves were identified as one of the most critical coastal habitats in the area. Some critical issues in relation to mangroves were as follows:
- Mangroves provide the primary source of income for two of the six Chwaka Bay pilot villages; for the remaining four villages, it provides a supplementary source of income;
- Although the mangrove forest has remained constant in size, the quality of mangrove trees (and, hence, the relative value of the harvested products) is declining;
- Poor law enforcement, lack of community participation, low carrying capacity of the resource, and lack of site-specific knowledge on mangrove ecology were among the major constraints to sustainable management;
- Potential opportunities were not yet recognized for economic non-consumptive uses of the mangrove forest;
- Neither consensus on the precise boundaries of village tenure nor the current management practices are built on the village tenure system.

Recommended strategy

The ICAM process found that, apart from developing some issue-specific strategies, a common institutional framework for the management was quite important. Therefore, the following were agreed:
- The Chwaka Bay–Paje pilot site should be non-legally designated as an integrated multisectoral planning area with specific boundaries;
- A multisectoral coastal resources management committee (CRMC) should be created under the chairmanship of the Regional Administrator to assist planning and implementation.

Since the inception of the ICAM strategies, implementation has started – although slowly, owing to shortage of funds. Beyond the Chwaka Bay–

Paje pilot site, the coastal profile for the whole of Unguja Island has been prepared.

Jozani–Chwaka Bay Conservation Project

In 1995, the Commission for Natural Resources (CNR) and the Tanzanian branch of CARE International (a development agency that operates 20 integrated conservation and development projects worldwide), initiated the Jozani–Chwaka Bay Conservation Project to conserve one of the critical biodiversity-rich areas of Zanzibar and improve community development of the surrounding communities. Community participation in the project has been a main aim.

Shehia forests and their role in conservation and sustainable management

The Shehia (one or more villages) is the smallest unit in the local government hierarchy (fig. 19.2). Under the 1996 Act, the Commission enters into an agreement with Shehias who want to manage their own Shehia forests, including mangroves. The usual approach is for the Shehia to elect a conservation committee that, on behalf of the larger community, develops the management agreement consisting of Shehia land-use plans, Shehia Forest boundaries by-laws, and restoration and protection measures. Eight villages in the project area are in the final stages of preparing their resource-use management plan for submission to the Commission. Once approved by the Commission, the area becomes a Community Forest Management Area under the new act.

Jozani Environmental Conservation Association (JECA)

The Jozani Environmental Conservation Association (JECA) is a community organization that has been established for the purpose of protecting the local resources and enhancing the socio-economic development of the people around the Jozani and Chwaka Bay area. This organization was previously an advisory committee, representing all eight project Shehias. In 1997, the advisory committee expressed its desire to develop into an independent local NGO for the purpose of fostering sustainability of conservation and development activities when external support is phased out. Through project support, the advisory committee was successfully registered as an NGO in May 1999.

The development of JECA is one of the crucial initiatives in the empowerment of local communities in Zanzibar regarding conservation and

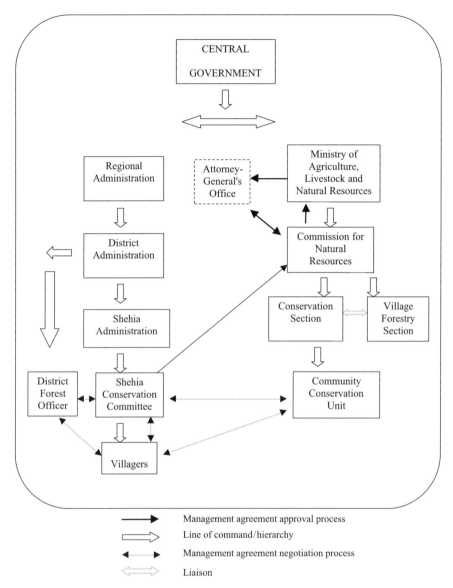

Figure 19.2 Institutional arrangements for community-based conservation in Zanzibar

development, which both CARE and the CNR are advocating. JECA has been involved in producing Shehia-level forest-management plans, forest patrols, and participatory research work. It is currently running a saving and credit scheme aiming at building people's assets to encourage alternatives to woodcutting. A total of 76 five-person loan groups have registered, with TSh. 11.3 million (US$14,125) issued to them as loans.

Given that JECA leadership stems from the grass roots, capacity building is crucial at this early stage. The Jozani–Chwaka Bay project provides a series of technical training schemes in the areas of business and loan management, skill improvement, forest management, NGO, and financial management and accounting.

Mangrove-planting activities

One of the recent management interventions to restore small, degraded mangrove sites is mangrove planting. This activity has never been implemented by the CNR directly, but always by communities. Schoolchildren at Kisakasaka were the first to initiate the programme, which then spread to the entire community. Within the Jozani Project area, mangrove planting is currently being practised at Pete/Jozani, Muungoni, and Michamvi. At Muungoni, individual villagers have planted mangrove for sustainable exploitation: about 5 ha have been planted since 1995. This figure reflects the relatively small areas degraded to the point of needing replanting; however, it also represents a sense of ownership of the mangroves, with up to 60 villagers gathering on a Saturday to plant.

Mangrove Boardwalk, eco-tourism, and alternative-income generation

In 1997, a walkway was built in an area of community mangrove forest, close to where visitors see one of Zanzibar's main tourist attractions, the endemic Zanzibar Red Colobus Monkey (*Piliocolobus kirkii*). The walkway is run as a partnership between JECA and the Commission and, thus far, has raised over $20,000, which has been invested in community projects such as nursery schools, a mosque, and village electrification. This has encouraged the village to conserve their areas of mangroves more actively.

Sub-sector analysis of woodcutting industry

The Jozani project has carried out a participatory analysis of the woodcutting industry from producer to consumer for two villages (Charawe and Ukongoroni). The study was aimed at understanding the woodcutting industry in the villages, given the rural dependence and the increasing demand from urban areas. About 64 and 39 per cent of Charawe and Ukongoroni adult males, respectively, are engaged in commercial woodcutting. As a consequence, it was discovered that Charawe and Ukon-

Table 19.2 Forest area and standing volumes for Charawe and Ukongoroni

Type	Charawe				Ukongoroni			
	Area (ha)	%	Standing vol. (m³)	%	Area (ha)	%	Standing vol. (m³)	%
Coral rag	2,058	67	58,597	61	3,835	78	122,905	76
Mangrove	1,027	33	37,882	39	1,089	22	38,611	24
Total	3,085		96,479		4,924		161,516	

Source: Ely et al. (2000).

goroni villages supply 15 per cent of the mangrove poles and 75 per cent of the dryland poles currently sold in Zanzibar Town (Ely et al. 1999). Based on the total growing stock of natural forest produce, mangroves provide a significant contribution for both villages (table 19.2).

The annual allowable cut was 3,290 m³ and 5,369 m³ for Charawe and Ukongoroni, respectively. The greater area of forest, high total annual increment, and lower current annual extraction rate for Ukongoroni indicates that woodcutters from this village will be able to harvest at current rates for a longer period (over 20 years) without further destruction. However, the reason for the lower harvest rate in Ukongoroni is that villagers have already exhausted that part of the forest most readily accessible from the village.

Outstanding issues

Significant efforts have been made to bring Zanzibar's mangroves under sustainable management. Within the last five years, initiatives including policy, law, planning, community tourism, and mangrove management have all borne fruit. However, despite these initiatives, the forces of degradation are still formidable. Despite a shift to community participation and empowerment, the commission remains in conflict with those sections of the community, often the poorest, which are entirely dependent on mangrove resources. The commission struggles to reconcile its own policy thrusts of resource conservation and poverty alleviation.

General recommendations

Approved Zanzibar mangrove zoning and management plan

The entire mangrove ecosystem in Zanzibar is declared as a forest reserve. Despite this, degradation continues, and land-use change to salt extraction and hotel development occurs. The classification of mangroves into different ecological types and identification of those of particular conservation importance (and ensuring that they are protected) are necessary, as is evaluation of the various use zones, particularly for community management and eco-tourism development (CNR 1997). Additionally, mangrove forests are not well incorporated in the land-use plan for Zanzibar, although they constitute about 15 per cent of the natural vegetation of the country. Our recommendations, therefore, are evaluation of the mangrove ecosystem, zoning, and the development of the National Mangrove Management Plan for incorporation into the Zanzibar Land-use Plan.

Extension of community-based management

Mangroves are socially and culturally important resources for the rural Zanzibar population (Nasser 1994). Therefore, it is extremely difficult (it might be said, impossible) to isolate communities in any management initiative to be established. Given that communities have expressed their willingness to participate and that demonstration projects are quite promising, we recommend the extension of community based management to the whole country.

Promotion of integrated coastal-area management

Zanzibar is currently planning a strategy for the expansion of the tourist industry and foreign investments to strengthen her economy. Most of this development, particularly the tourist hotels, are built along the coastal belt, whereas rural communities have traditionally been using these coasts. The integrated coastal-area management approach would be essential, not only to avoid conflict with local communities but also to ensure that coastal resources are not depleted as a result of unplanned development. Every kind of development will be required to respect system sustainability.

Specific recommendations

1. Place a moratorium on new clearing of mangroves for salt extraction until policy (including Environmental Impact Assessment) for this is in place and the effectiveness of existing salt extraction is evaluated.
2. Classify Zanzibar mangroves into different ecological types and ensure that each type is represented with the Zanzibar protected-area system.
3. Increase the capacity of the CNR to extend the coverage of community forest-management areas.

4. Streamline and institutionalize the process of establishing community-based forest and mangrove forest within the CNR.
5. Work more closely with poorer mangrove-felling groups to find solutions to sustainable mangrove felling and maintaining and improving livelihoods.
6. Public education campaign to make communities, the tourist industry, and other ministries aware of the importance of mangroves.
7. Increase the capacity of new and existing community institutions to increase the role in mangrove management, including long-term forestry courses for community foresters.
8. Develop systems for monitoring the progress of community-based mangrove management.
9. Establish a biosphere reserve in southern Unguja, including much of the mangroves in that area.

REFERENCES

CNR. 1997. *The Zanzibar Long-term Forestry Plan*. Zanzibar: Commission for Natural Resources; Vantaa, Finland: Forest and Park Services of Finland.

COLE. 1996. *Towards Integrated Management and Sustainable Development of Zanzibar's Coasts*. Zanzibar: Commission for Lands and Environment.

Ely, A., A. Omar, A. Basha, S. Fakih, and R. Wild. 1999. *A Participatory Study of the Wood Harvesting Industry of Charawe and Ukongoroni*. Zanzibar, Tanzania: Commission For Natural Resources, CARE-Tanzania, and UNESCO.

FAO. 1995. *Formulation of a National Forest Policy for Zanzibar*. Rome: FAO.

GOZ. 1996. *The National Forestry Policy of Zanzibar*. Revolutionary Government of Zanzibar.

Griffith, L. 1950. *Working Scheme for the Mangroves of Zanzibar*. Zanzibar: Government Printers.

Krebs, R. 1997. *The Mammals of Jozani*. Zanzibar: School of International Training.

Leskinen, J., V. Pohjonen, and M. Ali. 1997. *Woody Biomass Inventory of the Zanzibar Islands. Zanzibar Forestry Technical Paper No. 40*. Zanzibar: Commission for Natural Resources; Vantaa, Finland: Forest and Park Services of Finland.

Mohammed, S.M. 1999. *The Ecology and Socioeconomy of Chwaka-Bay*. Zanzibar: Commission for Natural Resources and CARE-Tanzania.

Nasser, S. 1994. *Socio-economic and Ecological Aspects of Mangrove Forest Management in Zanzibar*. Aas, Norway: Agriculture University of Norway.

Pakenham, R. 1984. *The Mammals of Zanzibar and Pemba Islands*. Zanzibar.

Shinula, J. 1990. *A Survey on the Distribution and Status of Mangrove Forests in Zanzibar, Tanzania*. Zanzibar Environmental Study Series No. 5. Zanzibar: Commission of Lands and Environment.

Summary of presentations and guidelines, and action plan

20
Summary of presentations and guidelines for future action

Marta Vannucci and Zafar Adeel

Almost a hundred participants attended the workshop. Lively discussions, exchanges of information, and suggestions followed every presentation. Mangroves have more than just regional significance: at present, most maritime nations with mangroves along their coasts are aware of their importance. The Asia-Pacific region has, area-wise, the largest amount of luxuriant, well-developed mangrove ecosystems. Much of the tropical, coastal wetlands include mangroves which, to date, have been affected by humans, to a greater or lesser extent, by pressures including shrimp farming, urban and industrial development, and tourism. Some types of use (or misuse), ranging from total destruction to reforestation and afforestation or re-afforestation – either failures or successes – have now been studied. Both failure and success stories are, as a rule, useful to estimate and quantify the significance and the role of mangroves in the functioning of the tropical coastal zone. The importance of mangroves is, at present, universally recognized; however, how best to deal with this unique type of ecosystem is still a problem that requires investigation, discussion, and collaboration among scientists, decision makers, and the coastal dwellers themselves.

The need for action to preserve and manage mangroves better was highlighted at the UNU-UNESCO-ISME International Workshop. This is also in accordance with Chapter 17 of Agenda 21, in which an urgent need for utilizing the region's coastal and marine resources in a sustainable manner is emphasized. The workshop proved to be a most suitable forum for interesting and fruitful interaction. It is not possible to summarize, even briefly, all the topics and suggestions that were made; only some of the most pressing problems are mentioned below.

The participants of the workshop discussed the various actions that can be taken collectively, institutionally, and individually. There is a need to conserve and protect the existing mangrove ecosystems in the face of economic and population pressures. At the same time, efforts need to be launched to rehabilitate those areas that have become deteriorated and degraded. In this respect, it is crucially important to raise the awareness of the general public, policy makers, and private concerns alike. This can be achieved by targeting the messages from this workshop to the appropriate audience in a relevant and focused manner. Exchange of information between experts on mangroves, governments, and the general public can help to identify crucial needs of the various areas for further development.

Thematic discussions

Focus was on the title theme – the role that science must play to face problems of health, growing human population density, nutrition, poverty alleviation, environmental conservation, natural-resources management, and the fundamental issue – which is that of acquiring a better understanding of the structure and function of the mangrove ecosystem itself. The knowledge acquired should also (in the words of the Director-General of UNESCO in his Opening Address to the Workshop) ensure that "sciences should help to overcome the problems that the world is facing also in terms of culture, tradition, societies, civilizations, and a changing world."

Among the many fundamentally important topics that were raised, the following are amongst those in greatest need of attention:
- The importance of forest-species diversity;
- Long-term comprehensive plans for the management of mangroves;
- Sustained monitoring, research, and evaluation of natural forests; of biosphere reserves; and of natural, managed, and man-made mangrove forests and plantations.

Some mangrove-management tools mentioned were legislation for mangrove conservation; integrated coastal-management practices; community participation and empowerment; legal backing for community participation; management agreements with community organizations as NGOs; and nominated and paid stewardship by selected households or individuals to supervise the enforcement of uses, rules, and regulations.

The concept of co-management or broad partnership arrangements for resources management between users, government, the community, national and international NGOs, and others was explained to some extent. Examples drawn from successful case studies in the Philippines and Mexico exemplified what problems may arise and some ways of solving them.

A realistic legislature and enforcement of the law are everywhere felt as a key issue at government level; however, in practice and in the long run, the community's role is essential to regulate tree cutting, fishing, resource management, and jurisdiction.

Attention was drawn to the fact that traditional laws and customs of certain nations already function along the same principle. Examples were drawn from Fiji; from certain Javanese and Sumatran communities in Indonesia; the Philippines; and Pakistan; and from others.

Generally felt difficulties are conflicting interests between the inhabitants and outsiders, as well as among different groups of people in the coastal zone itself – such as fishermen, woodcutters, cattle rangers, and newcomers – for instance, for building shrimp-ponds. Merits and demerits, as well as threats posed by large-scale commercial exploitation as opposed to small village-level traditional uses, were discussed. It was emphasized that more should be known about the site-specific production potential before decisions are taken and rules and regulations determined. Management practices should be revised at intervals of a few years.

Man-made forests and extensive plantations of forest species monitored over long periods of time and studied statistically allow for better-defined management practices.

Techniques for rational management and restoration of degraded mangrove ecosystems, developed at specific locations with success, were discussed with great interest.

Caution is recommended regarding the transfer of techniques and methods for the restoration and/or management of degraded ecosystems, because conditions of the wetlands and the wants and requirements of the people vary widely from place to place.

Many details regarding site-specific conservation and management were discussed, as set out in the specific chapters.

- There are challenges of all sorts, there are no universal solutions, and most problems are site specific and require special attention.
- The relationships between mangroves and both capture and captive fisheries vary from place to place; the uses made by humans have been either positive or negative but have seldom been sustained successfully for decade-long periods.
- Threats to the mangrove ecosystem, or to constituent elements of the

system, require the attention of policy makers and scientists, as well as the will and involvement of local communities for sustainable use and continued production.

- Practical actions to be taken include the following: identification of the needs of the people and the use to which the mangrove can be put; the taking of measures for the rehabilitation of the system; analysis of traditional procedures of use and management; and the securing of the collaboration of scientists, other professionals, funding agencies, government, and policy makers.

In conclusion, and quoting from the summaries of the presentations, some of the outstanding recommendations appear to be as follows: propagation of species for biodiversity; methods and techniques for management; international cooperation and exchange of scientists; monitoring and evaluation of mangrove programmes; "know-how" transfer for all aspects of mangrove management; case-study examples; scientific study of traditional knowledge; the constant bearing in mind that tidal wetlands, or mangroves, are a special ecosystem that must be treated as a whole, rather than the enhancement of production of single elements to a level of maximum yield per unit area or unit time. Solving the problems that refer to one element (or a few elements) of the system cannot be an answer to the general problems.

From the experience in Tanzania, where the mangroves of Zanzibar have been exploited for many centuries (probably since the time of Imperial Rome), we learn that some mangrove-management tools used successfully in Zanzibar are legislation for mangrove conservation, integrated coastal-management strategies, community participation and empowerment, legal backing for community participation, and management agreement with communities.

The points emphasized from the explosive, intensive, aquaculture experiences in the Philippines and Viet Nam teach us that the concept of co-management is essential for success and should involve a broad partnership between users, governments, NGOs, etc. This concept and practice of stewardship grants "Certificates of Stewardship" contracts to households, families, or individuals who, under contract, are paid to ensure that previously agreed, site-specific rules and regulations for the use of the mangrove are observed by everyone. Traditional knowledge should always be studied attentively.

Research should have (or lead to) an integrated holistic approach, including taking advantage of the results of natural or man-made calamities (such as the spraying of mangroves with defoliants in the Viet Nam War, and the destructive forces of cyclones).

Research – basic, mission oriented, practical, and site specific – was unanimously considered to be essential in the long run to solve routine and recurrent problems and also for solving occasional problems arising in daily managerial practice. Close association among scientists, other competent professionals, and the coastal people, is indispensable.

Although details and viewpoints necessarily varied among participants coming from so many different countries, there was a very high level of agreement on most issues.

These are some (but far from all) of the important questions raised. For most, there were no ready answers. However, we may confidently say that the success of the workshop was largely due to the fact that, with case-specific variations, all the participants agreed on the main issues. The workshop was an interesting step in the right direction. Our goals are similar, but the ways and means of reaching those goals are site specific. We all learn from each other's experience, acquired from both basic and applied research. Both types of research are needed to develop the wisdom needed for the management of that highly dynamic system – the tropical coastal zone.

21

Mangroves action plan

Zafar Adeel

Introduction

This chapter provides an overview of the many features recommended for action to conserve, protect, and rehabilitate mangrove ecosystems. The action items can be divided into three broad categories: (1) information dissemination and capacity building; (2) basic and applied research and development projects; and (3) development of sustainable-development systems.

Information dissemination and capacity building

The following action items related to information dissemination were identified by the workshop:
- Publication of proceedings of the workshop, including all the papers, summary of discussions of the technical sessions, and this Action Plan. The United Nations University (UNU) will undertake this task as part of the workshop activities. The final proceedings will be published both in hard-copy form and disseminated on the Internet.
- To increase the level of information exchange, it was suggested that Internet-based networking should be encouraged. The Global Mangrove Database and Information System (GLOMIS) mangroves listserver and the UNU listserver for coastal wetlands were suggested as possible options.

- It was suggested that information regarding mangrove-management techniques should be developed and disseminated as a set of guidelines.
- Public education and awareness raising – at various levels from community to international – should be a key point of dissemination activities; cooperation of the popular media should be sought in this respect. Inclusion of mangrove issues in formal education processes (especially those for children) should be a part of this process.

The following action items were suggested to undertake capacity building at various levels:
- There should be increased coordination between various international and regional institutions in undertaking capacity-building activities. Information dissemination and networking can greatly assist in this respect.
- There should be emphasis on training the educators and trainers for conservation of mangroves. The UNU biodiversity Training Workshop on Marine Biodiversity in Mangroves and Coastal Ecosystems was cited as an example and endorsed by the workshop participants as a regional activity. Subsequently, an Agreement of Cooperation has been achieved between UNU, UNESCO, and Annamalai University, India, to hold similar training programmes on an annual basis.

Applied research and development

The following gaps in the existing knowledge and research base were identified at the workshop:
- The impact of market forces and economic development on mangroves, particularly issues of sustainable development and socio-economic evaluation of mangrove ecosystems. These issues are closely linked to the livelihood of people dependent on mangroves.
- The knowledge and applicability of various legal and institutional aspects, particularly at national and international levels.

The following action items were identified for further action on applied research and development of mangroves:
- An in-depth evaluation of the introduction of foreign species of mangrove plants into coastal habitats. The need for a closer look at biotechnology issues was also highlighted.
- Development and evaluation of abandoned shrimp- and fish-ponds in areas adjacent to or overlapping mangroves.
- Quantitative evaluation of freshwater requirements for mangrove ecosystems.
- Investigations into the impacts of anthropogenic activities on various sub-ecosystems contained within the mangroves.

- Development of some quantifiable indicators for evaluation of mangrove ecosystems, including socio-economic factors involved. These can then be applied to some detailed case-study analyses.
- A closer look at the impact of chemicals and pollutants in the mangrove ecosystems.
- A detailed analysis of the traditional practices for utilization and conservation of mangroves. These can then be further improved through scientific investigation and/or newly available technologies.
- Evaluation of various novel uses of mangrove plants, including for medicinal purposes and as a food source.

A major action item suggested by the workshop participants was to develop an inventory of mangroves in the region. This should include the current status of mangroves, maps of mangroves in the region, the areas at further risk, highlights of major conservation activities, modes of community interaction, and a glossary of terms for mangroves. The UNU and World Resources Institute (WRI) agreed to team up to produce a document *Habitats at Risk*, along the lines suggested by the workshop participants. Such a document will help to raise awareness of the general issues relevant to mangroves and identify priorities for research and conservation activities. Indirect benefits would include assisting donor agencies in identifying critical areas needing immediate attention.

Sustainable management systems

The following action items were suggested as relevant to development of sustainable management systems for mangroves:
- Capitalizing on UNESCO's experience with biosphere reserves in developing similar reserves for mangroves. A tangible step in this direction was the workshop on Mangroves and Biosphere Reserves organized by UNESCO in the Pacific islands during the year 2001. This activity can also be coupled with the designation of selected mangroves as World Heritage Sites.
- Development of methodologies and processes for conflict resolution in the context of conservation and development of mangroves. This should focus primarily on multi-stakeholder interests.
- Development and evaluation of ecosystem goods and services that mangroves can provide. Such activities will help to ensure sustainability of the mangroves in the long term.
- Investment in the mangroves to achieve economic security, stable livelihood, food security, and reduction of poverty. A major emphasis should be on development of communities reliant on the mangroves for their livelihood.

- Development of eco-tourism along the lines of some examples presented at the workshop. Again, this would help in sustainable management of mangroves and assist in improving the livelihood of communities dependent on mangroves.
- Development of links with various multilateral environmental agreements and conventions is important.

Acronyms

ACTMANG	Action for Mangrove Reforestation
ADB	Asian Development Bank
ASPACO	Asia Pacific Cooperation for Sustainable Use of Renewable Resources in Biosphere Reserves and Similarly Managed Areas
BFAR	Bureau of Fisheries and Aquatic Resources
BFRI	Bangladesh Forest Research Institute
CBO	community-based organization
CNR	Commission for Natural Resources
CSC	Certificate of Stewardship Contracts
DBH	diameter (at) breast height
DENR	Department of Environment and Natural Resources
DGRLR	Directorate-General of Reforestation and Land Rehabilitation (currently the Directorate-General of Reforestation and Land Rehabilitation and Social Forestry)
EIA	Environmental Impact Assessment
EIS	Environmental Impact Statement
FADs	fish aggregating devices
FLAs	Fish-pond Lease Agreements
FRECMA	Forestry Resources Conservation and Management Act
GLOMIS	Global Mangrove Database and Information System
IBRD	International Bank for Reconstruction and Development
ICAM	Integrated Coastal Area Management
ICMAM	Integrated Coastal Marine Area Management
ICZM	integrated coastal-zone management
ISME	International Society for Mangrove Ecosystems

ITTO	International Tropical Timber Organization
IUCN	International Union for Conservation of Nature and Natural Resources
JAM	Japan Association for Mangroves
JECA	Jozani Environmental Conservation Association
JICA	Japan International Cooperation Agency
MAB	Man and Biosphere (UNESCO programme)
MBEPF	Management Board for Environmental Protection Forests
MERC	Mangrove Ecosystem Research Center
MFA	mangrove-friendly aquaculture
MNR	Ministry of Natural Resources
MRCRMP	Mangrove Rehabilitation and Coastal-resource Management Project
MSSRF	M.S. Swaminathan Research Foundation
NCC	Non-Commercial Cover
NGO	non-governmental organization
NPO	non-profitable organization
NPV	net present value
ppt	parts per thousand
RFD	Royal Forest Department
RRDP	Rainfed Resources Development Project
SEAFDEC	Southeast Asian Fisheries Development Center
SFD	Sindh Forest Department
TAC	Total Allowable Catch
UNESCO	United Nations Educational, Scientific and Cultural Organization
UNU	The United Nations University
USAID	US Agency for International Development

List of contributors

Zafar Adeel Assistant Director (program Development), United Nations University, International Network on Water, Environment and Health. He has experience in a variety of environmental issues, including water-pollution problems, water-resource management in dry areas, solutions to industrial environmental problems, modelling of environmental systems, and environmental policy development. By training, Dr Adeel is an environmental engineer with postgraduate degrees from Carnegie Mellon University and Iowa State University.

Kazuhiro Ajiki was born in Yamagata, Japan in 1962. He finished the Graduate School of Science, Tohoku University, and now is an associate professor of the Faculty of Humanities and Social Sciences, Mie University. He is studying human geography, especially interested in socio-economic structure and environmental problems in rural areas of Japan and South-East Asia.

A.F.M. Akhtaruzzaman, a research scientist in forest utilization, obtained the degree of Doctor of Technology from Helsinki University of Technology in Finland. He holds the position of the Director, Bangladesh Forest Research Institute, the only national institute for forestry research in the country. Dr. Zaman has over 33 years of research and management expertise, and authored 60 research publications in leading international and national journals. He acted as the Editor of Bangladesh Journal of Forest Science, an internationally recognized journal. He also taught applied chemistry in the University of Chittagong. He supervised Master's and Ph.D. theses in different Universities in home and abroad. Dr. Zaman visited many countries to discharge his professional responsibilities.

Sanit Aksornkoae is now working as Professor Emeritus at the Faculty of Forestry, Kasetsart University, Bangkok, Thailand. He gained a PhD degree in Plant Ecology, particularly mangroves, from Michigan State University, USA in 1975. He has taught and researched on mangrove ecology since 1972 after gaining an MS degree from the Faculty of Forestry, Kasetsart University, Thailand. Professor Sanit Aksornkoae was elected to several national and international committees on mangroves and coastal resources. He is now the Treasurer of the International Society for Mangrove Ecosystems.

Shigeyuki Baba has a Doctor's degree in Agriculture and worked in Forest Science at Faculty of Agriculture, University of the Ryukyus which is the only national university in Okinawa Prefecture, Japan. As Associate Professor, he lectures on Silviculture, Tree Genetics and Small-Island-Biology, etc. When the International Society for Mangrove Ecosystems (ISME) was established in 1990, he was designated as Deputy Executive Secretary of ISME. Since 1996, he has become Executive Secretary of ISME in Voluntary base. He has planed and implemented many mangrove projects as Executive Secretary of ISME, and participated many scientific meetings on mangroves.

Zhongyi Chen is a reseacher of South China Institute of Botany (SCIB), Chinese Academy of Sciences, Guangzhou,China. He served as Deputy Director of SCIB from 1998–2000, and as a professor of the Graduate School of the Chinese Academy of Sciences, Beijing and South China Agriculture University, Guangzhou of China. He specializes in cytotaxonomy and works in the Department of Taxonomy of SCIB.

Dr. Sonjai Havanond has Ph.D in the field of Bioregulation from Tokyo University of Agriculture. He has worked in Thai mangrove research and development about 25 years. His position now is Coastal and Mangrove Resources Research and Management Expert in Department of Marine and Coastal Resources, Ministry of Natural Resources and Environment.

Atsuo IDA is Chief Advisor to "the Project on Capacity Building for Forestry Sector in Cambodia" by Japan International Cooperation Agency (JICA) and Department of Forestry and Wildlife since December 2001. He served as Team Leader to "the Development of Sustainable Mangrove Management Project in Indonesia" by JICA and Ministry of Forestry from 1997–99. He is originally a senior staff of Forestry Agency Japan

Shinichiro Kakuma is an Executive Researcher in Research Institute for Subtropics, based in Okinawa Japan. He specializes in co-management of coastal fisheries resources and fish aggregating devices in the tropics and subtropics.

Brenda M. Katon is currently a consultant of the Asian Development Bank-Operations Evaluation Department in Manila, Philippines and a PhD student in public administration and governance at the University of the Philippines. In the past, she was a consultant of the International Center for Living Aquatic Resources Management (ICLARM).

Dr. Fujimoto Kiyoshi is an associate professor of Faculty of Policy Studies, Nanzan University, Japan. He is an environmental geographer, especially in geomorphology, and has studied on mangrove environment since 1987. He was engaged in the mangrove studies as a researcher (1991–1995) and senior researcher (1995–2000) of the Forestry and Forest Products Research Institute, Ministry of Agriculture, Forestry and Fisheries of Japanese Government.

Kiyomi Kogo is an Associate Professor of Tohoku Gakuin University. She has experience in a variety of environmental issues, including socio-economic study on the utilization of mangrove forests in Vietnam and Ecuador. She has Master of Science degree from University of Tokyo.

Motohiko Kogo is an adventurer of the mangroves. He has received several prizes by his activities on mangroves, including Rolex Awards for Enterprise, Nikkei Award for Technology of Global Environmental Issues, Minamata Award for Environments, etc. For this ten years, he is challenging to conserve and restore mangrove ecosystems in Vietnam, Ecuador and Myanmar.

Thabit S. Masoud is the Assistant Area Coordinator for CARE International in Tanzania, (Zanzibar). He served as head of Integrated Conservation and Development (ICD) for the Commission of Natural Resources in Zanzibar between 1994–January 2002 and as Project Manager for Jozani-Chwaka Bay Conservation Area, a partnership initiative between CARE and the Government of Zanzibar. Professionally, he is a conservation biologist.

Zebin Miao is a research assistant of South China Institute of Botany (SCIB), Chinese Academy of Sciences, Guangzhou, China. He specializes mangroves conservation in south China.

Reiko Minagawa was born in Yokohama, Kanagawa Prefecture, Japan in 1958. In 1981 she graduated in the Faculty of Agriculture, Tokyo University of Agriculture, and in 1994 she completed a graduate PhD Programme in Bioregulation Studies in that university. In 1995 she became an assistant and in 2000 a part-time instructor, in Tokyo University of Agriculture. Her specialties are mangroves, epiphytes and vines, and forest ecology.

Patricia Moreno-Casasola has a PhD in plant ecology from the University of Uppsala and has also undertaken a degree in Sustainable Development in FLACAM. She is currently a researcher at the Institute of Ecology A.C. in Mexico. She lectures in the graduate programs of this institution and coordinates the Management Plan for La Mancha Watershed in the Gulf of Mexico.

Takehisa Nakamura was born in Nagano, Japan, in 1932. In 1956 he became a graduate of the Faculty of Agriculture, Tokyo University of Agriculture, and in 1964 he became a Doctor of Science of Tohoku University. From 1986 until his retirement in 2002 he was Professor of the Faculty of Agriculture of Tokyo University of Agriculture, and he is now an Emeritus Professor of that university. His specialties are in mangroves, Pteridophyta, and forest ecology.

Phan Nguyen Hong has been involved in mangrove ecosystem research for more than 35 years. After the reunification of the country, he has been a leader of several national and international projects on the mangrove ecosystem in Viet Nam. He is the author of books in Vietnamese and English. He has chaired (and edited a number of proceedings of) national workshops on mangrove and coastal wetland conservation. Phan Nguyen Hong holds the degrees of BSc (1956), MSc (1964), PhD (1970), and DSc (1991) in ecology. He is at present Professor of Ecology at Hanoi University of Education and Director of Mangrove Ecosystem Research Division, Centre for Natural Resources and Environmental Studies, Viet Nam National University, Hanoi.

Robert Pomeroy is an Associate Professor-in-Residence and Sea Grant Extension Specialist, Department of Agricultural and Resource Economics, University of Connecticut-Avery Point, Groton, Connecticut, USA. He holds a PhD in Natural Resource Economics. His main areas of interest are policy analysis, fisheries management and development, aquaculture economics, coastal-resource management, and international development. He was a Senior Scientist at the International Center for Living Aquatic Resources Management, The World Fisheries Center.

Jurgenne H. Primavera has a Ph.D. in Marine and joined the Aquaculture Department of the Southeast Asian Fisheries Development Center (SEAFDEC AQD) based in Iloilo, central Philippines. As Senior Scientist, her research first focused on penaeid shrimp aquaculture, in particular, broodstock development and extensive grow-out systems, and more recently shifted to the shrimp-mangrove fisheries linkage and the development of integrated mangrove-aquaculture. She has written numerous scientific articles, review papers, manuals, book chapters, technical reports and other publications, and attended various scientific conferences, symposia and workshops on aquaculture, fisheries, mangroves, and the environment.

M. Tahir Qureshi has a MS in Forestry and MS is Parasitology. He has worked as a senior technocrat in Forest Department, Government of Pakistan. He is currently working in IUCN as Programme Director in Coastal and Marine Ecosystem. He is the author of the Management Plan of the Indus Delta.

Mesake T. Senibulu is currently the government's Divisional Surveyor Central/Eastern of the Department of Lands and Surveys, Ministry of Lands and Mineral Resources. He has a Bachelor of Arts degree in Land Management and Development from the University of the South Pacific, Fiji and also a Diploma for Graduates in Spatial Information Studies from the University of Otago, New Zealand. He had served as co-coordinator of foreshore development applications for four years from 1992 and within this period attended the first JICA and ISME sponsored course in Sustainable Management of Mangrove Ecosystems, conducted at Okinawa, Japan. Mr. Senibulu is a registered land surveyor

Dr. Shigemitsu Shokita is a professor and councilor of the University of the Ryukyus, Okinawa, Japan. He makes a special study of the taxonomy, life-history and aquaculture of crustaceans such as shrimp and crab, and studies on the role and function of mangrove ecosystem at the Ryukyu Islands.

AN Subramanian is a Professor at the Centre of Advanced Study in Marine Biology of Annamalai University, India, where he instructs masters and doctoral students in Marine Sciences. His doctoral research in India was on the pollution status of Pichavaram Mangrove, India. Subsequently, he obtained another PhD from Ehime University, Japan for his work on persistent pollutants in penguins and marine mammals. He is a member of several collaborative programmes with Ehime University, Japan; Otsuchi Marine Research Centre; Tokyo University, Japan; and United Nations University, Tokyo, Japan. He is the organizer of the UNU–UNESCO International Training Course on Mangroves held annually at the Centre of Advanced Study in Marine Biology, Annamalai University, India. He has been awarded fellowships by the Ministry of Education, Culture, Sports, Science and Technology, Japan; and the Japan Society for Promotion of Sciences, Japan for carrying out research work in Japan. His present field of interest is the global transport of persistent pollutants in the terrestrial, coastal, and open-ocean environments, for which he is a member of several national and international projects. He has been instrumental in the production of seven PhDs, has been the author or co-author of about 70 national and international publications, and has edited two manuals.

Aprilani Soegiarto has a Ph.D. in Marine Science from University of Hawaii, Honolulu, Hawaii. At present he is the President of ISME (International Society for Mangrove Ecosystem) and Chairman of the Board of Trustees PROSEA (Plant Resources of South East Asia). He is a research professor in oceanology at the Indonesian Institute of Science (LIPI) in Jakarta and professor at the Bogor Agriculture University, Bogor.

Dr. M. Vannucci is the Former Chief Technical Adviser of UNDP/UNESCO Mangrove Project for Asia and Pacific implemented during 1983–88. She has PhD from the Univ. of S. Paulo, and is specialized in Biological Oceanography. She has more than 100 scientific papers to her credit. She is a member of the Brazilian Academy of Science of Brazil since early 1950's. In 1997, she was granted the Grand Cross of the Order of Scientific Merit, which only selected scientists receives. She is Former Vice-President (1990–1999) and Acting President (1999) and an Honorable Technical Adviser to ISME since 1999.

Ruijiang Wang is an assistant professor of South China Institute of Botany, the Academy of Sciences, Guangzhou, China. He is a PhD candidate in the Department of Ecology and Biodiversity, The University of Hong Kong (2000–2003). He is interested in the vascular plants systematics and evolution basing on the morphology and molecular techniques.

Robert G. Wild is Co-Management Adviser (UK DFID) to the Protected Area Department of the Turks and Caicos Islands. From 1989–2000 he worked in East Africa on integrated conservation and development projects for CARE and WWF. He is an ecologist with social science qualifications. He is a member of the Institute of Ecology and Environmental Management, UK.

Index

Page references followed by *t* indicates a table; followed by *fig* indicates an illustrated figure; followed by *p* indicates a photograph.

Acharya, G., 122, 123
ACIPHIL, Inc., 223, 224
Ackerman, 162
ACTMANG (Action for Mangrove Reforestation) [Japan], 124
ACTMANG project (Viet Nam)
  education and awareness as part of, 241, 242*t*–243*t*
  goal of, 234
  reforestation during, 236
  research during, 234–236
  sketch of Viet Nam areas in, 238*fig*
  sustainable management for conservation in, 236–237*t*
  sustainable use of mangrove ecosystem, 239–240
Adams, T.J.H., 208, 211
Adeel, Z., 297, 302
Agbayani, R.F., 204
Agena, M., 91, 99, 100
Agenda 21, 297
Agreement of Cooperation (UNU, UNESCO, and Annamalai University), 303
Ajiki, K., 196, 257

Akhtaruzzaman, A.F.M., 249
Aksornkoae, S., 149, 156, 264
Alcala, A.C., 194
Alcott, M., 23
Ali, M., 281
Alongi, D.M., 120
Alonzo-Pasicolan, S.N., 196
Amphibians
  Can Gio District
    in Mangrove Biosphere Reserve, 127, 132
  reforestation effects on, 121–122
Anon, 60, 63, 121, 126
Aquaculture development
  Crustacean species
    found in Okinawan mangals, 105*p*, 106*p*
    found in Okukubi River, 92*t*–94*t*
  experiments
    chemical analysis of mangrove leaves, 96–97
    description of first, 95, 96*t*
    description of second, 95–96
    feeding habiits of estuarine fishes, 99*fig*–100

preference of *Helice leachi* for brown leaves, 97–99
fish species found in Okinawan mangals, 109*p*, 110*p*
food chains
  in mangroves of Iriomote Island, 101*fig*
  and role of aquatic animals in degradation of mangrove litter, 100, 102
Indonesia, mangrove planting and traditional, 276*t*–277*t*, 278–279
introduction to, 76
MFA (Mangrove-Friendly Aquaculture) categorizations, 204
molluscs found in Okinawan and Thai mangals, 107*p*
Philippines, guidelines to mangrove zones as sites for ponds, 201*fig*
research aims, objectives, and findings on, 77, 79–95
  comparison of species composition/ abundance of tree fauna, 90*t*
  distribution/abundance of macrobenthos, 79–80
  feeding habits of crabs, 91, 95
  feeding habits of *Helice* species, 95
  feeding habits of molluscs, 89, 91
  macrofaunal community, 78*t*
  macrofaunal habitats and dominant species, 85, 87
  mangal-dwelling taxa, 77, 78*t*, 79
  results at Ranong mangal, 82–85, 86*fig*
  results at Smare Kaow mangal, 80, 81*fig*, 82
  stock enhancement of economically important species, 89
  Thailand study sites, 80*fig*
  tree-dwelling fauna, 87–89
Thailand mangrove forests, 157
tree-dwelling fauna on Ranong mangal, 108*p*
Asano, T., 234
ASEAN (Association of South-East Asian Nations), 55
ASPCO (Asia Pacific Cooperation for Sustainable Use of Renewable Resources in Biosphere Reserve and Similar Managed Land), 6
Avicennia alba research methods/results, 42, 44

Avicennia marina
  in Bali, 29*p*
  in Galapagos Islands, 23*p*
  in Indonesia, 10, 21*p*
  in Miani Hor, 180*p*

Baba, S., 8, 144, 233
Baconguis, S.R., 192, 196
Balaji, V., 60
Balasubramanian, T., 70
Baling, N., 223
Bangladesh
  introduction to, 249
  mangrove forest research
    achievements of, 255–256
    BFRI mandate on, 253, 255
    need for, 253–255
  mangrove forests
    description of, 249–250
    economic importance of, 250, 252*t*
    map showing locations of, 251*fig*
    problems of, 250, 253
Ban Tha Pradu (Viet Nam), 265, 266*t*–267*t*
Below-ground carbon sequestration study
  listed in Asia-Pacific region, 141*t*–142*t*
  methods used in, 139–140
  relationship between carbon-burial period/stored carbon, 144*fig*
  results and discussion, 140, 143
  review of past studies, 138
Berry, A.J., 85, 91
BFAR (Bureau of Fisheries and Aquatic Resources), 192, 224, 225
BFRI (Bangladesh Forest Research Institute), 253, 255
Bhatia, O., 77
Bhosale, L.J., 60
Birds
  Can Gio District
    in Mangrove Biosphere Reserve, 133–134
    reforestation effects on, 122
Blasco, F., 63, 66
Blatter, E.J., 60
Blok, L.G., 194
Bohnsack, J.A., 213
Bosire, J., 7
Brackish-water communities (Can Gio District), 114–115
Briggs, S.V., 138
Brown, W.H., 196

Bruguiera gymnorrhiza (Okinawa), 27*p*, 28*p*
Bryson, J.M., 185
Buot, I.E., Jr., 192

Cabahug, D.M., Jr., 196
Camilleri, J., 91
Can Gio District (Viet Nam)
 below-ground carbon sequestration study in, 139–140, 143
 Duyen Hai Forestry Enterprise (1978) established in, 116
 Duyen Hai Forestry Enterprise, 116, 117
 historical development of mangroves in, 113–117, 119
 lodges established for education and research, 248*p*
 Mangrove Biosphere Reserve
  amphibians and reptiles in, 132
  fish in, 130–131
  mangrove flora in, 127
  zoobenthos in, 128–129
 mangrove forest/forestry units in
  listed, 117*t*
  map of, 118*fig*
  socio-economic study of utilization of, 261–262*t*–264
 map of, 112*fig*
 MBEPF (Management Board for Environmental Protection Forests), 117, 119
 natural conditions of, 112–113
 reforestation
  effects of, 119–123
  environmental changes due to, 123–124, 125*t*
  establishment of program, 111–112*fig*
  rare animals and, 123*t*
 See also Ho Chi Minh City (Viet Nam)
Can Gio Forestry Park, 122
Caratini, C., 63, 66
CARE International, 288, 289
Centre for Advanced Studies (India), 6
Champion, H.A., 60
Chapman, V.J., 123, 257
Charcoal kiln (Thailand), 26*p*
Charuppat, J., 151
Charuppat, T., 151
Chen, R.H., 138
Chen, Z., 45
Chokoria Sundarbans (Bangladesh), 253
Christensen, B., 138

Chwaka Bay
 mangrove forests of
  ICAM programme for, 287–288
  Jozani-Chwaka Bay conservation project, 288–291*t*
  management, conservation, development of, 286–287
  outstanding issues of, 291
 See also Zanzibar
Clarke, C.B., 60
CNR (Commission for Natural Resources) [Zanzibar], 284, 288, 290
Cogtong Bay Central Visayas Regional Project, 223
Cogtong Bay mangrove rehabilitation
 brief history of resource mangement, 222–223
 changes in resource management, 223–225
 characteristics of functional co-management, 228–229
 co-management approach used in, 219–220
 description of, 220–222, 221*fig*
 incentives to cooperate, 225–226
 outcomes of co-management arrangements, 226–228
Cogtong Bay RRDP (Rainfed Resources Development Project), 223
Conservation. *See* Mangrove conservation
Cooke, T., 60
Crosby, B.C., 185
Crustacean species (Okukubi river), 92*t*–94*t*
Cuong, V.V., 113, 116, 119

Dai, A.Y., 95
Dalzell, P., 211
Daniel, P.A., 91, 98
Darsidi, A., 52
Dat, H.D., 121
Davie, J.D.S., 203
Davie, P.K.F., 89
Day, J., 138
Day, J.W., Jr., 79
DENR (Department of Environment and Natural Resources) [Cogtong Bay], 223, 224, 225, 258
Deshmukh, S., 60
DGRLR (Directorate-General of Reforestation and Land Rehabilitation), 270

INDEX

Dickson, J.O., 216
Dolar, M.L., 194
Duyen Hai Forestry Enterprise (Viet Nam), 116, 117
Dyes extracted from mangrove trees (Okinawa), 27*p*

Fa'asili, U., 208, 209, 212, 213
FADs (fish aggregating devices), 209, 216
FAO (Food and Agriculture Organization), 212
Fiji
  description of, 161–162
  government protections
    conservation efforts of, 163–164
    Department of Lands and Surveys and, 164–165
  land tenure classifications in, 163
  mangrove forests
    distribution and extent of, 162
    sustainable harvesting of, 165–166
    types and associated flora, 162–163*t*
Fiji National Mangrove Management Committee, 164
Fiji National Mangrove Management Plan, 164
Fiji Native Land Trust Board, 163
Fischer, A.F., 196
Fish/fisheries
  Can Gio District
    in Mangrove Biosphere Reserve, 130–131
    reforestation effects on, 121
  overexploitation of, 6–7
  Pakistan mangrove rehabilitation and, 176, 178
  Philippines, effects of fish-pond construction, 260–261
  tropical/subtropical co-management approach to
    catch transition of fish, 209*fig*
    catch transition of shellfish/sea urchin, 210*fig*
    conditions in, 210–211
    gill-net fishery in Iriomote Island, 216–217*t*
    management tools and, 211–213
    Samoa and Okinawa comparative case study on, 213*fig*–216
    as solution to problems, 208
    state of the resources, 209–210

FLAs (Fishpond Lease Agreements), 224
Flora of British India (Hooker), 60
Flores, E.C., 208
Forests. *See* Mangrove forests
Forest Survey of India, 61–62
Frith, J.W., 77
Fujimoto, K., 138, 139, 140

Gamble, J.S., 60
Gervain, P., 216
Giddens, R.L., 91
Giesen, W., 53
Gill-net fishery (Iriomote Island), 216–217*t*
GLOMIS (Global Mangrove Database and Information System), 13, 302
GLOMIS project website, 6
Golley, F.B., 138
Gopal, B., 60, 62
Griffith, L., 281, 282
Guangdong Province (China) project
  background information on, 45, 47
  development of plantation, 49
  effect of mangrove introduction during, 49–50
  height of *Sonneratia apetala* after planting, 49*t*
  mangrove species of, 46*t*
  mean height of *Sonneratia apetala*, 49*fig*
  reforestation efforts of, 47–48
Gunderman, N., 77

Hamazaki, K., 89
Hargis, T., 138
Havanond, S., 39, 204
Heald, E.J., 91
Hegerl, E.J., 203
Heritiera littoralis (Okinawa), 10, 16*fig*, 18*p*
Higa, Y., 212
Hinrichsen, S., 181
Hirugi. *See* Mangroves
Ho Chi Minh City Decision No. 173 CT (1991), 116–117
Ho Chi Minh City Resolution No. 165/QD-UB (1978), 116
Ho Chi Minh City (Viet Nam)
  map of, 112*fig*
  reforestation programme undertaken by, 111–112
  See also Can Gio District (Viet Nam)
Hong, P.G., 233
Hong, P.N., 111, 115, 121, 234

Hooker, J.D., 60
Hortus Bengalensis (Indian report, 1814), 60
Hussain, Z., 122, 123
Hvding, E., 208, 211

ICAM (Integrated Coastal Area Management) programme (Zanzibar), 281–288
ICMAM (Integrated Coastal Marine Area Management) [India], 72
Ida, A., 270
India
  Centre of Advanced Study in Marine Biology, 70, 72
  Department of Ocean Development of the Government of India, 72
  mangrove ecosystems in
    areas under mangroves, 61t
    chemical studies in, 67, 70
    eastern seaboard, 63–64
    history of research on, 59–61
    Pichavaram mangrove forest, 64–66, 65fig, 68t 69t, 71fig, 72fig
    statewise distribution (1999) of, 62fig
    Status Report (1987) on, 61–62
    wealth of fauna in, 66–67
    western seaboard, 63
  variable ecological features of, 59
Indian Pichavaram mangrove forest
  described, 64–66
  physical and chemical parameters observed in, 68t–69t, 71fig, 72fig
  on the south-east coast, 65fig
Die Indo-Malaysichen Strandflora (Schimper report, 1891), 60
Indonesia
  Avicennia marina in, 10, 21p
  mangrove ecosystems in
    brief description of, 52–54
    conservation research on, 51–57
  Mangrove plantations in, 32p, 33p
  shrimp ponds in, 20p, 28p
  Sonneratia sp. in, 19p
  sustainable mangrove management project
    handbook for mangroves, 274
    implementation of, 271
    introduction to, 270
    mangrove planting/aquaculture, 32p, 276t–277t, 278–279

  Nursery Manual summary of published outcomes, 272–274, 273fig
  objectives of, 270–271
  products, 271–272fig, 276t–277t
  silviculture manual for, 274, 275fig
  sustainable management models, 275, 278
  tambak-parit system of land use in, 51
Indonesian conservation research
  background information on, 51–52
  brief description of mangrove ecosystems, 52–54
  history of research programmes, 54–56
  management and conservations efforts, 56–57
Indonesian Department of Forestry and Crop Estates, 56
Indonesian National Mangrove Committee, 54–55
INEGI-SEMARNAP, 183, 184
Iriomote Island
  below-ground carbon sequestration study in, 139, 140, 143
  gill-net fishery in, 216–217t
  mangroves, 17p
ISME (International Society for Mangrove Ecosystems), 3, 12–13fig, 55, 60
ITTO/ISME project, 6
IUCN (International Union for the Conservation of Nature), 176, 177
Izumo, K., 234

Janiola, E., 223
Japanese prefectures, 8–9
JECA (Jozani Environmental Conservation Association), 288–290
JICA (Japan International Cooperation Agency), 7, 10, 270
Johannes, R.E., 208, 211
Jones, D.M., 77, 91

Kakuma, S., 208, 212, 216
Kannan, L., 60
Kannan, R., 65, 66
Kannupandi, T., 65, 66
Kartawinata, K., 52, 53
Karthikeyan, E., 70
Kathiresan, K., 60, 65, 72
Katon, B.M., 219
Kawasaki, K., 212
Keenan, C.P., 89

Kelekolo, L., 212
Kelly, P.F., 196
Kikuchi, T., 139
King, M., 208, 209, 213
Kisakasaka Village (Zanzibar), 284–286
Kobashi, D., 236, 240
Kogo, K., 233, 234
Kogo, M., 233
Krebs, R., 281
Krishnamurthy, K., 60, 62
Kubo, H., 91
Kusmana, C., 138

Lagin, A., 216
Lal, P.N., 166
Lee, S.Y., 91, 138
Leskinen, J., 281
Limsakul, S., 91, 100
Lugo, A.E., 138

MAB (Man and Biosphere) [UNESCO], 3, 126
McManus, J.W., 210
MacNae, W., 52, 77, 89, 91
Magi, M., 235, 236, 240
Mammals
  Can Gio District
    rare animals in, 123t
    reforestation effects on, 122–123
Mancrofaunal community (South-East Asian and Okinawa), 78t
Mandal, R., 62
Mangrove conservation
  Chwaka Bay, 286–287
  Indonesian research/efforts on, 51–57
  ISME role in, 13fig
  Mexican coastal-management project and issue of, 183–184
  NGOs (non-governmental organizations) role in, 12–13
  Pakistan, 176–178, 177t
  Philippine regulations on, 202t
  Thailand, policy formed for, 149–151
  UNU-UNESCO-ISME International Workshop on
    described, 5, 297–298
    thematic discussions during, 298–301
  Zanzibar, initiatives for, 284–286, 291–293
Mangrove conversion
  Pakistan, 168–174

  Philippines regulations on fish-ponds and, 200
Mangrove ecosystems
  degradation and destruction of, 8
  described, 8, 9, 11–14
  establishing goals for, 12fig
  importance of, 9–11
  Indian, 59–72fig
  Indo-Malaysian evolutionary origins of, 59
  Indonesian conservation research on, 51–57
  Japanese public perceptions on, 9
  role of aquatic animals in
    experiments on, 95–102
    introduction to, 76
    research aims, objectives, and findings on, 77–95
  roles of ISME in conserving, 13fig
  unique properties of, 151
Mangrove forests
  below-ground carbon sequestration study
    listed in Asia-Pacific region, 141t–142t
    methods used in, 139–140
    relationship between carbon-burial period/stored carbon, 144fig
    results and discussion, 140, 143
    review of past studies, 138
  degradation and destruction of, 8, 19p
  Indonesian sustainable management of, 270–279
  of Okinawa, 17p
  overexploitation of, 6–7
  Pichavaram mangrove forest (India), 64–66, 65fig, 68t–69t, 71fig
  South-East Asia
    introduction to, 257
    Philippines case study (1992), 258–259t, 260–261
    Thailand case study (1996-1997), 264–265
    Viet Nam case study (1994), 261, 262t–263t, 264
  Thailand
    area of land-use zones by province (1998), 154t
    changes in existing, 152t
    policy and management practices for, 153–157
    policy for research and applied knowledge, 157–158

Mangrove forests (cont.)
  sketch map showing main areas, 159*fig*
  status and quality of, 151, 153
Mangrove nursery (Bali), 31*p*
Mangrove plantations
  Bali experimental, 31*p*
  Bangladesh, 253
  Guangdong Province (China) project, 49
  Indonesia, 32*p*, 33*p*
  Thailand, 35*p*, 156
  Viet Nam, 34*p*
Mangrove planting
  Fiji, 33*p*
  height of *Sonneratia apetala* after, 49*t*
  Indonesia, 32*p*, 276*t*–277*t*, 278–279
Mangrove reforestation
  ACTMANG project (Viet Nam), 236, 247*p*
  Can Gio District
    effects of, 119–123
    environment changes due to, 123–124, 125*t*
    programme established, 111–112*fig*
    rare animals and, 123*t*
  Thailand policies and practices for, 156
Mangrove rehabilitation
  Cogtong Bay
    brief history of resource mangement, 222–223
    changes in resource management, 223–225
    characteristics of functional co-management, 228–229
    co-management approach used in, 219–220
    description of, 220–222, 221*fig*
    incentives to cooperate, 225–226
    outcomes of co-management arrangements, 226–228
Mangrove research
  on aquaculture development, 66–95
  Bangladesh, 249–256
  Can Gio District lodges established for, 248*p*
  history of Indian, 59–61
  Indonesian conservation, 51–57
  mangroves action plan on, 303–304
  Sonneratia alba, 40–43*fig*
  Thailand policies and practices for, 157–158
Mangrove Research Centre (Ranong Province), 158

Mangroves action plan
  applied research and development, 303–304
  information dissemination and capacity building, 302–303
  sustainable management systems, 304–305
Mangroves and Biosphere Reserves workshop, 304
Mangroves of India (report, 1987), 60, 61
Mangroves (Iriomote Island, Okinawa), 17*p*
Mangrove timber, 25*p*, 26*p*
Mangrove trees (Florida), 22*p*
Mangrove trees (Okinawa), 16*p*, 22*p*
Mangrove trees (Viet Nam), 23*p*
Mann, D.L., 89
Mapalo, A.M., 192
Marte, C.L., 208
Masoud, T.S., 280
Mazda, M., 235, 236, 240
MBEPF (Management Board for Environmental Protection Forests) [Viet Nam], 117, 119
Mehra, R., 223
MERD (Mangrove Ecosystem Research Division), 126, 234
Mexican coastal-management project
  communities located in area of, 184–186
  conserving and managing mangroves, 186
  description of, 182–183
  ecosystems ecology programmes, 190–191
  education programmes, 188–189
  environmental legislation/programmes, 187–188
  local economy programmes, 189
  politics and associated programmes, 189–190
  problem of mangrove conservation within, 183–184
  programmes designed to help society, 186–187
  stakeholders in area of, 185*t*
MFA (Mangrove-Friendly Aquaculture) categories, 204
Mia, M.Y., 91, 95
Miao, Z., 45
Mien, P.V., 121
Minagawa, R., 39
Miyagi, T., 138, 143, 234
Miyake, S., 95
Miyamoto, C., 240

INDEX   321

Mochida, Y., 139
Mohammed, S.M., 280
Moreno-Casasola, P., 181
MRCRMP (Mangrove Rehabilitation and Coastal-resource Management Project), 227
MSSRF (M.S. Swaminathan Research Foundation), 70, 72
Munro, J.L., 210
My, T.V., 115

Nakamura, T., 39
Nakasone, Y., 77, 91, 99, 100
Nam, V.N., 113, 115
Naskar, K., 62
Nasser, S., 282, 283, 292
Natividad, A.C., 216
Nayak, S.T., 61
NEQA (National Environmental Quality Standards), 171
Network Foundation, 224
NGOs (non-governmental organizations)
  concept of co-management with, 299
  mangrove conservations efforts by, 12–13
  report on local community support for reforestation by, 234
  resource management incentives for, 225
Nipa palm leaves (Viet Nam), 24p, 25p
Nishihira, M., 91
Nishimura, K., 240
Noor, Yus Rusila, 53
Nuique, J., 194
Nursery Manual (Indonesia), 272–273fig

Odum, H.T., 138
Odum, W.E., 91, 98
Okinawa
  Bruguiera gymnorrhiza, 27p, 28p
  crustacean species found in, 105p–106p
  fishery comparative study on Samoa and, 213fig–216fig, 214fig, 215fig
  fish species found in, 109p, 110p
  Heritiera littoralis, 10, 16fig, 18p
  Iriomote Island
    below-ground carbon sequestration study in, 139, 140, 143
    gill-net fishery in, 216–217t
    mangroves, 17p
  Mancrofaunal community, 78t
  mangrove forests, 17p
  mangrove trees
    dyes extracted from, 27p

    photographs of, 16p, 22p
  molluscs found in, 107p
  Rhizophora stylosa, 27p, 28p
Okinawa Development Agency, 209
"Okinawa Mangrove Fund" (proposed), 14
Okinawa Prefecture, 3, 8
Okinawa-Samoa fishery comparative study, 213fig–216fig, 214fig, 215fig
Okukubi River Crustacean species, 92t–94t

Pakenham, R., 281
Pakistan
  Avicennia marina plantation in, 180p
  estimated number of camels (1996) in, 175t
  Indus delta
    mangrove rehabilitation along, 177t
    Rhizophora mucronata in, 180p
  Indus River
    average annual discharge, 169t
    percentage salinity in, 170t
  mangrove conversion
    causes and consequences of, 168–174
    into salt-pans, 174
    urbanization and industrialization impact on, 170–171
  mangrove forests
    condition and composition of, 168
    use for fodder, 172t
    monthly consumption for fuel, 173t
    overexploitation of, 171, 174
    species listed, 169t
  mangrove rehabilitation
    coastal community development case study, 176–178
    in Indus delta/Balochistan coast, 177t
  mangrove resources
    conflicting interest in use of, 174
    management plans for sustainable use of, 174, 176
"Perum Inhutani" (Indonesia), 56
"Perum Perhutani" (State Forest Corporation) [Indonesia], 56
Philippines
  Fisheries Decree of 1975, 196
  mangrove/brackish-water culture-pond area
    changes in, 198fig
    description of, 197t
    guidelines to mangrove zones as sites for, 201fig
    parallel trends in, 199fig

322 INDEX

Philippines (cont.)
  mangrove conservation, regulations on, 202*t*
  mangrove conversion
    guidelines on sites for aquaculture ponds, 201*fig*
    regulations on fish-ponds and, 200
  mangrove forests
    functions and valuation of, 194, 195*t*, 196
    major and minor species of, 193*t*
    socio-economic study of utilization of, 258–259*t*, 260–261
    status of, 192, 194
    threats to, 196–197, 203
Pichavaram mangrove forest (India), 64–66, 65*fig*, 66*t*–67*t*, 71*fig*
Pinto, L., 192
Plaziat, J.C., 89, 91
Pohjonen, V., 281
Pohnape Island (Micronesia), 35*p*
Pohnpei Island below-ground carbon sequestration study, 139, 140
Polunin, N.V.C., 211
Pomeroy, R.S., 209, 219, 227
Popper, D.M., 77
Por, F.D., 77
Prain, D., 60
Primavera, J.H., 192, 196, 203, 204

Qureshi, M.T., 167
Quy, N.D., 113

Rajendran, B., 70
Ramesh, A., 70
Reptiles
  Can Gio District
    in Mangrove Biosphere Reserve, 132
    reforestation effects on, 121–122
Research. *See* Mangrove research
Restoration. *See* Mangrove reforestation
Rhizophora apiculata (Bali), 30*p*
Rhizophora mucronata (Bali), 30*p*
Rhizophora sp. (Ecuador), 18*p*
Rhizophora stylosa (Okinawa), 27*p*, 28*p*
Richmond, 162
Robertson, A.I., 91, 98
Ronnback, P., 194
Ross, P., 115
Ruddle, K., 208, 211

Saenger, P., 162, 203
Saintilan, N., 138
Sakai, T., 95
Samoa-Okinawa fishery comparative study, 213*fig*–216*fig*, 214*fig*, 215*fig*
Sang, H.T., 245
Sasekumar, A., 77, 120
Schatz, R.E., 196
Scholl, D.W., 138
Scott, D.A., 164
SEAFDEC Aquaculture Department, 204
Senibulu, M., 161
Shinula, J., 281
Shokita, S., 76, 77, 87, 88, 89, 91, 99, 100, 101
Shrimp farming (Viet Nam), 239–240, 247*p*
"Shrimp Fever" (1980s), 196
Shrimp ponds
  Bali, 31*p*
  Ecuador, 20*p*
  Indonesia, 20*p*, 28*p*
Siddiqi, N.A., 250
Sidhu, S.S., 63
Silvicultural system
  Indonesian practice of, 274, 275*fig*
  Thailand practice of, 155–156
Singh, V.P., 63
Snedaker, S.C., 138
Soegiarto, A., 51, 52, 53
Soemodihardjo, S., 52, 54
Sonneratia alba
  described, 39
  seeds of, 29*p*
Sonneratia alba research methods/results
  on age, cable rotts, and pneumatophores, 40
  comparison of proportions of pneutomatophores, 43*fig*
  on correlation between young tree hight/cable-root thickness, 43*fig*
  discussion and conclusions, 44
  on numbers and types of pneumatophores, 41*fig*
  on proportions of various types of pneumatophores, 43*fig*
  on shape of pneumatophores and evironment, 40–42
Sonneratia sp.
  introduction to Gunangdong Province (China), 45–50
  photographed in Indonesia, 19*p*

South Pacific Commission, 164
Stephenson, A., 85
Stephenson, T.A., 85
Subramanian, A.N., 59, 70
Suzuki, S., 234

TAC (Total Allowable Catch) system, 212
Takeoka, H., 70
Tang, V.T., 121
Tantanasiriwong, R., 77
Taquet, M., 216
Thailand
  aquatic study sites in, 80*fig*
  below-ground carbon sequestration study in, 139, 143
  mangrove forest conservation
    measure for land use in areas of, 154–155
    policy formed for, 149–151
    policy and management practices for, 158
  mangrove forest reforestation, 156
  mangrove forests
    aquaculture development in, 157
    area of land-use zones by province (1998), 154*t*
    changes in existing, 152*t*
    policy and management practices of, 153–157
    policy for research and applied knowledge, 157–158
    silvicultural system practices for, 155–156
    sketch map showing main areas of, 159*fig*
    socio-economic study of utilization of, 264–265
    status and quality of, 151, 153
  tree-dwelling fauna on Ranong mangal, 108*p*
Thailand RFD (Royal Forest Department), 150, 158, 264
Thanikaimoni, G., 63, 66
Tirsrisook, K., 91, 100
Tissot, C., 63, 66
Tomlinson, P.B., 192
Tri, N.H., 115
Tsuchiya, M., 91
Tsuruda, K., 234
Tuan, L.D., 116
Twilley, R.R., 91, 138

UNDP/UNESCO Regional Mangrove Projects, 176
UNEP (United Nations Environment Programme), 287
UNESCO's *Bibliography on Mangrove Research 1600-1976*, 234
UNESCO (United Nations Educational, Scientific, and Cultural Organization), 3, 12, 55, 186, 240, 304
United Nations University, 12
University of the Ryukyus, 3
Untawale, A.G., 60, 61
UNU Biodiversity Training workshop on Marine Biodiversity in Mangroves and Coastal Ecosystems, 303
UNU-UNESCO-ISME International Workshop
  described, 5, 297–298
  thematic discussions during, 298–301
UNU (United Nations University), 3, 302, 304

Vannucci, M., 3, 60, 297
Van Steenis, C.G.G.J., 52
Vega, M.J.M., 226
Viet Nam
  below-ground carbon sequestration study in, 139, 143
  Can Gio District
    below-ground carbon sequestration study in, 139–140
    Duyen Hai Forestry Enterprise, 116, 117
    historical development of mangroves in, 113–117, 119
    lodges established for education/research in, 248*p*
    Mangrove Biosphere Reserve, 127–132
    mangrove forest/forestry units map, 118*fig*
    mangrove forestry units listed, 117*t*
    map of, 112*fig*
    MBEPF of, 117, 119
    natural conditions of, 112–113
    reforestation effects, 119–124, 123*t*, 125*t*
    reforestation programme of, 111–112*fig*
    socio-economic study of utilization of mangrove forests in, 261–262*t*–264

Viet Nam (cont.)
   mangrove conservation (ACTMANG project)
      education and awareness as part of, 241, 242t–243t
      goal of, 234
      introduction to, 233–234
      reforestation, 236, 247p
      research on, 234–236
      sketch of project area, 238fig
      sustainable management for, 236–237t, 239
      sustainable use of mangrove ecosystem, 239–241
   mangrove plantations in, 34p
   mangrove trees, 23p
   Nipa palm leaves in, 24p, 25p
   shrimp farming in, 239–240, 247p
Viet Nam War, 115–116
Viet, V.Q., 120

Wang, Ruijang, 45
Warner, G.F., 77, 91
Watling, D., 162, 163
Wild, R.G., 280
Williams, M.J., 209
Wilson, A.F., 138
Women's Union (Viet Nam), 236
Woodroffe, C.D., 138
WRI (World Resources Institute), 304

Yang, S.L., 95

Zanzibar
   CNR (Commission for Natural Resources) of, 283
   Forest Reserve Decree of, 283
   introduction to, 280–281
   mangrove forests of
      conservation initiatives for Kisakasaka Village, 284–286
      described, 281–282fig
      history of management of, 282–283
      improved policy and legislation on, 286
      institutional arrangements for conservation of, 289fig
      recommendations for, 291–293
      threats to management of, 283–284
   See also Chwaka Bay

VMC - framework (55)
conservation efforts - Indonesia (56)
industries (54)
biodiversity (53)
what Indonesia has done - 52+
India - early m. mgment (61)
fauna - food sources (66)   fishponds (56)
weather patterns (66)
chem. studies (67)
tourism - "wonderlands" puts pressure on the ecosystem (71)
satellites (71), uses   - successes (72)
marine ranching (89)   role of aquatic animals (89)
Food web (100). * feeding grounds (102).
Viet Nam - sprayed w/ herbicides (111) reforestation program
   - excellent results (115) war (116) after war
   2 phases of reforestation (116) (121) - wildlife

Catalogue Request

Name: _____

Address: _____

_____

Tel: _____

Fax: _____

E-mail: _____

To receive a catalogue of UNU Press publications kindly photocopy this form and send or fax it back to us with your details. You can also e-mail us this information. Please put "Mailing List" in the subject line.

53-70, Jingumae 5-chome
Shibuya-ku Tokyo 150-8925, Japan
Tel: +81-3-3499-2811 Fax: +81-3-3406-7345
E-mail: sales@hq.unu.edu   http://www.unu.edu